Expanding Our Understanding of the Psychosocial Work Environment:

A Compendium of Measures of Discrimination, Harassment and Work-Family Issues

Developed for the NORA Special Populations at Risk Team by the
Center for Women & Work, University of Massachusetts Lowell

Meg A. Bond
Alketa Kalaja
Pia Markkanen
Dianne Cazeca
Sivan Daniel
Lana Tsurikova
Laura Punnett

NIOSH Editor
Sherry Baron

Technical Support
Fang Gong
Donna Pfirman

DEPARTMENT OF HEALTH AND HUMAN SERVICES
Centers for Disease Control and Prevention
National Institute for Occupational Safety and Health

Disclaimer

Mention of any company or product does not constitute endorsement by the National Institute for Occupational Safety and Health (NIOSH). In addition, citations to Web sites external to NIOSH do not constitute NIOSH endorsement of the sponsoring organizations or their programs or products. Furthermore, NIOSH is not responsible for the content of these Web sites.

This document is in the public domain and may be freely copied or reprinted.

Ordering Information

To receive documents or other information about occupational safety and health topics, contact NIOSH at

Telephone: **1–800–CDC–INFO** (1–800–232–4636)
TTY: 1–888–232–6348
E-mail: cdcinfo@cdc.gov

or visit the NIOSH Web site at **www.cdc.gov/niosh**.

For a monthly update on news at NIOSH, subscribe to NIOSH *eNews* by visiting www.cdc.gov/niosh/eNews.

DHHS (NIOSH) Publication No. 2008–104

December 2007

SAFER • HEALTHIER • PEOPLE™

Foreword

In 1996, NIOSH created the National Occupational Research Agenda to advance occupational safety and health research for the nation. This agenda encompassed 21 priority research areas, including Special Populations at Risk. This priority area was created in recognition of the fact that the nation's increasingly diverse workforce contains many women, older workers, and racial and ethnic minorities. Disparities in the burden of disease, disability, and death are experienced by these groups, due in part to their disproportionate employment in high hazard industries and to certain social, cultural and political factors.

In order to advance the national research agenda, NIOSH partnered with the National Institutes of Health to fund pioneering new research to better characterize the role of environmental and occupational exposures in the development of health disparities for these populations.

The NIOSH grantee under this research initiative was the University of Massachusetts Lowell. Some of the important risk factors explored as part of this research were the role of workplace discrimination, harassment and work-family issues in the occurrence of occupational injuries and illnesses. While there is an increasing body of scientific evidence demonstrating the contribution of psychosocial stressors such as discrimination on health, these researchers found that the body of prior occupational safety and health research exploring them was limited. Occupational health studies that examine these factors will contribute to a better understanding of their role in causing or exacerbating health problems.

However, in the past, the limited availability and lack of awareness of appropriate methods of measurement of these potential workplace stressors has been a barrier. This document was developed by the investigators from the University of Massachusetts Lowell at the request of the Special Populations at Risk Team to fill that gap by disseminating to the broader occupational safety and health community a concise and accessible compendium of measures used by health researchers to assess the following domains:

- racism and racial/ethnic prejudice
- sexism and sexual harassment
- gender and racial discrimination
- work-family integration and balance
- support for diversity in the workplace/workforce

The issues, terms, and concepts addressed in the peer-reviewed studies that are cited and summarized in this document have profound emotional impact for people, individually and collectively. The nature of the document is such that the authors have to use sensitive terms and concepts frankly, so that the measures are meaningful and the document can fulfill its purpose as a research tool. While there is need for expanded research into the potential role of these stressors in the occurrence of occupational injuries and illnesses, many of the scales included in this compendium may be incomplete or inadequately tested in diverse work environments. It is NIOSH's hope that making these existing measures available will assist occupational safety and health researchers in the design of studies that further contribute to our understanding of their role and encourage further development of improved methods for occupational safety and health research to address this important gap. With improved understanding of the role of these stressors, occupational safety and health practitioners can also more successfully design and measure the impact of workplace intervention programs.

John Howard, M.D.
Director, National Institute for
Occupational Safety and Health
Centers for Disease Control and Prevention

Table of Contents

Foreword iii

Introduction 1
   Expanding our Understanding of the Psychosocial Work Environment 3
      Discrimination, Harassment, Workplace Biases, and Health 4
         Race-Related Dynamics 4
         Gender Dynamics 5
         Work-Life Integration and Health 6
         Sexual Orientation 7
         Interacting Influences 8
      Measures Described in This Compendium 9
      Validity and Reliability: Definitions Applied in This Document 12

Summary of Measures 13
   General Diversity Measures, Diversity Climate, Multiple ISMs 15
      Race, Racism, Ethnicity, Racial Discrimination & Related Measures 18
      Sexism & Sex discrimination 23
      Sexual Harassment 26
      Work-Family/Work-life Related Measures 28
      Sexual Orientation/Homophobia 36

General Diversity Measures, Diversity Climate, Multiple Isms 37
   Homophobia and Racism Scales 39
   Modified Godfrey-Richman ISM Scale (M-GRISMS-M) 41
   Perceived Supervisory Discrmination 44
   Diversity Climate 46
   Workforce Diversity Questionnaire (WDQ) 49
   Quick Discrimination Index 53

Race, Racism, Ethnicity, Racial Discrimination & Related Measures 59
   Institutional Racism Scale (IRS) 61
   Hispanic Stress Inventory (HSI) 66
   Perceptions of Racism Scale (PRS) 72
   The Racism and Life Experience Scales (RaLES) 75
   Workplace Racial Bias 84
   Krieger Measure of Experiences of Discrimination 87
   Schedule of Racist Events (SRE) 90
   Modern Racism Scale 93
   Perceived Racism Scale 97
   Motivation to Respond without Prejudice 100
   Acculturative Stress Scale (ACS) 105
   Cultural Mistrust Inventory (CMI) 108
   Racism Reaction Scale (RRS) 111
   Index of Race-Related Stress (IRRS) 113
   Race-Related Stress 118

Table of Contents

Sexism & Sex Discrimination 121
    Sexist Attitudes Toward Women Scale (SATWS) 123
    The Ambivalent Sexism Inventory (ASI) 127
    The Ambivalence toward Men Inventory (AMI) 132
    Schedule of Sexist Events (SSE) 135
    Stigma Consciousness Questionnaire (SCQ) 139
    Working Environment For Women In Academic Settings 146
    Working Environment for Women in Corporate Settings 149
    Old-Fashioned And Modern Sexism Scale 153
    Everyday Sexism 156
    General Attitudes toward Affirmative Action (AA) and Men's Collective Interest (CI) 159
    Neosexism Scale 162

Sexual Harassment 165
    Responses to Sexual Harassment and Satisfaction with the Outcome 167
    Sexual Experiences Questionnaire - Latinas (SEQ-L) 170
    Sexual Experience Questionnaire (SEQ-W) 173
    Organizational Tolerance for Sexual Harassment Inventory (OTSHI) 177
    Sexual Harassment Inventory (SHI) 180

Work Family/Work-Life Measures 185
    Work-Home Conflict 187
    Parental After-School Stress (PASS) 189
    Parental After-School Stress (PASS) 190
    Work Schedule Fit 192
    Informal Work Accommodations to Family (IWAF) 195
    Job-Family Role Strain Scale 199
    Family Management Scale 203
    Spillover Between Home and Job Responsibilities 207
    Work-Family Policies-Perceived Management Support and Usability 209
    Employer Support for Family 212
    Work-Family Interference and Tradeoffs 214
    Adjustment of Work Schedule 217
    Work-Family Conflict 219
    Survey Work-Home Interaction-Nijmegen (SWING) 222
    Work-Family Conflict 227
    Work-Family Interference 230
    Interrole Conflict Scale 233
    Work-Family Conflict and Family-Work Conflict Scales 237
    Worker Perception of Work Spillover 241
    Work-to-Family Conflict 244

Sexual Orientation: Heterosexism & Homophobia 247
    Discrimination Based on Sexual Orientation Questionnaire 249
    Heterosexual Attitudes Toward Homosexuals (HATH) 251
    Workplace Heterosexist Experiences Questionnaire (WHEQ) 254

References 247

Introduction

Introduction

EXPANDING OUR UNDERSTANDING OF THE PSYCHOSOCIAL WORK ENVIRONMENT
Meg A. Bond and Laura Punnett

There is broad recognition that the psychosocial environment at work can affect physical and mental health as well as organizational outcomes such as work performance and effectiveness. There is a substantial literature linking "job strain" and cardiovascular disease (Karasek & Theorell, 1990; Schnall, 1994; Belkic, Landsbergis, Schnall & Baker, 2004). The economic costs of job strain and job stress in general are related to absenteeism, turnover, and lost productivity, and, although difficult to estimate, could be as high as several hundred billion dollars per year (Karasek & Theorell, 1990). Thus for social as well as economic reasons, research aimed at understanding the conditions of work that contribute to physical and mental health concerns is well worth an intensified focus.

The psychosocial domains studied by occupational health researchers typically include psychological job demands, job control (decision latitude), social support, and intrinsic and extrinsic rewards (e.g., Karasek & Theorell, 1990; Siegrist, 1996). These factors, reflecting the organization of the work process, are often used to define the "psychosocial work environment." However, health and well-being are also affected by other features of the psychosocial work climate, such as unfair or inequitable treatment of employees, sexual harassment, and discrimination. Differential treatment, whether in terms of gender, age, race/ethnicity, sexual orientation, or disabilities, is increasingly recognized as a chronic stressor that can affect both psychological and physical health (Clark, Anderson, Clark, & Williams, 1999; Kessler, Mickelson, & Williams, 1999; Schulz, Israel, Williams, Parker, Becker, Becker, & James, 2000; Krieger, 2003). Experiences of discrimination can operate either in a cumulative way or in combination with each other (Swim, Hyers, Cohen & Ferguson, 2001; Essed, 1991). Furthermore, they are inherently likely to be distributed differentially by socioeconomic position (Kessler, Mickelson, & Williams, 1999).

Although it appears that discrimination experienced by members of target social groups has detrimental consequences, conceptual approaches and strength of findings vary, methodological problems with the literature have been noted (Meyer, 2003; Piotrkowski, 1997; Williams, Neighbors, & Jackson, 2003), and the evidence regarding *long-term* health outcomes is limited to date. Direct links to "upstream" organizational practices (e.g., workplace policies, programs, climate) have rarely been made empirically. Relevant literature is explored in more detail below, to summarize both our knowledge to date and the gaps in the empirical research, as well as to motivate inclusion of these work environment features in future studies. One barrier to such research is the lack of awareness of appropriate measurement instruments (Meyer, 2003). Thus the primary purpose of the current project has been to identify measures of gender and race-related dynamics in the workplace and to make them more easily accessible. Following the brief introduction and literature summary, this document catalogues 46 measures of biases, discrimination, and harassment that may be useful to occupational health researchers who wish to explore these issues further.

Introduction

DISCRIMINATION, HARASSMENT, WORKPLACE BIASES, AND HEALTH

Race-Related Dynamics

Racism occurs on many levels, from the interpersonal to the institutional. It has been defined as "an ideology of superiority that categorizes and ranks various groups, negative attitudes and beliefs about outgroups, and differential treatment of outgroups by individuals and societal institutions" (Williams, Yu, Jackson & Anderson, 1997, p. 338). At work, it can manifest in stereotypes and pigeonholing attitudes and assumptions, blocked opportunities, and limited access to resources needed to do one's work well. In addition, researchers are increasingly recognizing a more subtle form of racism, termed *aversive racism* (Dovidio & Gaertner 1996; Gaertner & Dovidio 1986), which involves underlying racially biased attitudes and behaviors of people who may not even be aware that their actions might be discriminatory. Aversive racism describes the scenario where people do not directly express more negative feelings about minorities or have lower expectations of any specific racial or ethnic group members; rather, they express fewer positive reactions to minorities and tend to favor majority group members (e.g., white men continue to receive more positive evaluations when all else is considered equal, Messick & Mackie 1989).

An increasing number of scholars are investigating the impact of racism, specifically in terms of its association with psychological well-being (Harrell, 1997; Klonoff, Landrine & Ullman, 1999; Neighbors, Jackson, Broman, & Thompson, 1996; Utsey & Ponterotto, 1996) and physical health (Jackson, Brown, Williams, Torres, Sellers, & Brown, 1996; Krieger, 2003; Krieger & Sidney, 1996; Rowley, 1994; Williams et al., 1997). The experience of racism is often both cumulative (i.e., daily and repeated) and additive across a variety of settings, such as the workplace, academia, and public places (Essed, 1991). Because of this pervasiveness and continuity over time, racism has been recognized as a chronic stressor (Utsey & Ponterotto, 1996; Green, 1995; Landrine & Klonoff, 1996). Increasingly scholars have identified racism as accounting directly for some of the differences in psychological and physical health between whites and people of color (Clark, et al., 1999; Landrine & Klonoff, 1996; Krieger, Rowley, Herman, Avery, & Phillips, 1993; Nazroo, 2003; Williams, et al., 1997; Utsey, Chae, Brown, & Kelly, 2002). For example, a thirteen-year panel study conducted by Jackson and colleagues (1996) demonstrated that experiences and perceptions of racial discrimination affected both the physical and mental health of African Americans. Krieger and Sidney (1996) showed that experiences of racial discrimination, as well as acceptance of unfair treatment as inevitable, were associated with higher levels of blood pressure in African American participants. However, much of this literature is cross-sectional, meaning that interpretation of the findings should proceed with the caveat that the directionality of the associations remains ambiguous.

The literature is still rather sparse on the specific health effects of racial discrimination in the work environment. Mays, Coleman, & Jackson (1996) examined the impact of perceived race-based discrimination on labor force participation and job-related stress among African American women. They found that perceived racism in the labor market affected advancement, skill development, and interpersonal relationships with co-workers. Similarly, Hughes and Dodge (1997) found that both interpersonal and institutional racism at work, especially interpersonal prejudice, were significant predictors of job satisfaction.

Introduction

People of color are often relegated to jobs with less control, high stress, and low influence (Blau, Ferber & Winkler, 2002). In addition, the racial make-up of the workplace can play a role, and some studies have explored how employees' racial background is associated with perceptions and experiences of the work climate. White employees often display a sort of blindness to racial dynamics and racist events, while people of color are more keenly aware of inequities and report higher levels of racial discrimination than whites (Watts & Carter, 1991; Weber & Higginbotham, 1997). Thus, it is perhaps not surprising that racial bias has been more frequently reported by people of color employed in predominantly white work settings (Hughes & Dodge, 1997).

Gender Dynamics

Like racial dynamics, gender-related stereotypes, prejudice, and discrimination operate as distinct sources of occupational stress (Korabik, McDonald, & Rosin, 1993; Swanson, 2000). Interpersonal manifestations range from sexist and racist jokes, demeaning comments, and harassment to team dynamics of avoidance and exclusion as well as lower expectations about women's competence and performance (Gutek, 2001; Swim et al., 2001; Pogrebin & Poole, 1997). Women's experience of sexist treatment – e.g., discrimination, negative sex stereotyping, isolation, and sexual objectification -- has been associated with mental heath concerns such as depression, anxiety, somatization and low self esteem (Klonoff, Landrine & Campbell, 2000; Landrine, Klonoff, Gibbs, Manning, & Lund, 1995; Swim et al., 2001) as well as with lower levels of physical health (including high blood pressure, ulcers, tension, and sleeplessness) (Nelson, Quick, & Hitt, 1989; Goldenhar, Swanson, Hurrell, Ruder & Deddens, 1998; Klonoff, Landrine & Campbell, 2000; Pavalko, 2003). Women who reported gender discrimination in their workplaces were found to have lower levels of job satisfaction and organizational commitment, as well as more negative relations with co-workers and supervisors, than those who did not experience gender discrimination (Murrell, Olson, & Hanson-Frieze, 1995). With regard to work outcomes, perceived sexism is associated with lower expectations and career aspirations and consequent choices for women (Evans & Herr, 1991). Furthermore, there is evidence that the negative impact of these gender-related stressors on health and well-being is above and beyond the effects of general job stressors such as overload (Swanson, 2000).

A related workplace stressor that affects the lives of many women employees is sexual harassment. The effects of sexual harassment in terms of work and health outcomes are similar to those of other forms of gender discrimination that occur in the workplace. Considerable research has found that sexual harassment is associated with negative psychological outcomes for women such as anxiety, depression, alienation, lower self-esteem, tension, and nervousness (Barling et al, 1996; Lenhart, 1996; Parker & Griffin, 2002), and negative somatic outcomes such as gastrointestinal disturbances, nausea, headaches, and insomnia (Gutek & Koss, 1993; Dansky & Kilpatrick, 1997; Goldenhar et al., 1998; Gutek & Done, 2001; Hesson-McInnis & Fitzgerald, 1997; Piotrkowski 1998). The experience of sexual harassment in the workplace has also been positively correlated with smoking and alcohol abuse (Richman, et al., 1999). In terms of work outcomes, sexual harassment was associated with loss of work motivation and higher levels of distraction that ultimately led to poor work performance, absenteeism, lateness, and turnover (Barling et al., 1996; Hanisch, 1996; Schneider, Swan, & Fitzgerald, 1997). Glomb and colleagues (1997) investigated the impact of indirect exposure to sexual harassment – i.e., being aware of negative treatment of women at work – and found that even harassment

Introduction

directed at someone else was associated with lower job satisfaction and increased work and job withdrawal, as well as with symptoms of psychological distress and somatization.

Sexual harassment is problematic not only in its own right but also because it seems to coexist with (stem from and/or result in) other gendered manifestations of negative work climates (Bond, 2003; Shrier, 1996; Gutek & Koss, 1993; Bingham & Scherer, 1993). For example, sexual harassment is more likely to occur in work climates characterized by high levels of sexist stereotypes and attitudes, and "everyday sexism" increases concerns about future provocation (Fitzgerald & Omerod, 1993; Murrell, Olsen & Hanson-Frieze, 1995; Pogrebin & Poole, 1997; Swim et al, 2001). Sexual harassment has been found to be more common in "sexualized" work environments and contexts where sex between colleagues is tolerated or condoned (Bond, 1995; Gutek, 2001). Fitzgerald and associates found that perceptions of an organization's responsiveness to employee concerns about harassment affected the frequency of sexual harassment (Fitzgerald, Drasgow, Hulin, Gelfand & Magley, 1997; Hesson-McInnis & Fitzgerald, 1997).

Gendered patterns also show up in the considerable occupational segregation that occurs across occupations and within general occupational categories (Blau et al., 2002; Wooton, 1997). *De facto* occupational segregation is not necessarily related to overt gender or race discrimination in a given workplace. Nevertheless, it is of great interest as a potential predictor of health status because "men's jobs" and "women's jobs" often have qualitatively and quantitatively different occupational exposures, whether physical work load, psychosocial strain, or even chemical exposures (Hall, 1992; Messing, 1995, 1997; Punnett & Herbert, 2000; Quinn, Woskie, & Rosenberg, 2000). Psychological job demands, decision latitude ("job control"), social support, and rewards affect both men and women, but they are unevenly distributed in the working population. For instance, jobs in which women predominate generally have lower decision latitude, on average, than men's jobs (Josephson et al., 1999; Karasek & Theorell, 1990; Matthews, Hertzman, Ostry, & Power, 1998; Nordander et al., 1999; Vermeulen & Mustard, 2000). There is evidence that the "job gender context" of work, conceptualized as the gender ratio of the workgroup and the gender traditionality of the work role, is related to the likelihood of experiencing sexual harassment, which, in turn, is subsequently associated with lower job satisfaction and psychological distress (Fitzgerald, Drasgow, Hulin, Gelfand & Maglev, 1997). In light of these findings, an intriguing observation is the report of increased sick leave for all causes, by both men and women, in jobs with high gender segregation – with the most problematic outcomes being for females in male-dominated groups (Alexanderson, Leijon, Akerlind, Rydh, & Bjurulk, 1994). It remains to be clarified whether gender segregation acts as a stressor *per se*, possibly as a source of psychosocial strain. Alternatively, it could be a determinant of, confounded by, or a proxy for gender differences in physical working conditions or other exposures, including harassment and discrimination.

Work-Life Integration and Health

Over the last couple of decades, there has been a dramatic increase in the labor force participation of married women with young children (Cohen & Bianchi, 1999; Hayghe & Bianchi, 1994) and in the percentage of married couples that are dual-earner families (Blau et al, 2002). The changing nature of the workforce has meant that an increasing portion of the workforce is facing the burden of combining work and family responsibilities, and many feel the stress.

Introduction

Research has indicated that about 40% of employed parents experience some level of conflict between their job demands and the demands of family life (Galinsky, Bond & Friedman, 1993). Although women in heterosexual marriages still take on the majority of the family-related responsibilities even when both spouses work (Blau, 1998), men's average weekly hours of housework have increased over the last 25 years (Blau, 1998; Bianchi, Milkie, Sayer & Robinson, 2000), and findings generally indicate that both mothers and fathers are affected by work-family conflict (Barnett & Brennan, 1995). For instance, Burden and Googins (1987) found that 36% of the fathers and 37% of the mothers in dual-wage families reported "a lot of stress" in balancing work and family responsibilities. Additionally, as more and more people's work schedules diverge from the traditional one of five 8-hour shifts per week, issues such as involuntary overtime and the spillover of work demands into unpaid, supposedly leisure time are intensifying conflict between work and family or personal life. Nevertheless, since women continue to have primary responsibility for childcare arrangements in many families, irregular schedules and involuntary overtime would be likely to cause particular problems for women workers (Büssing, 1996).

Health outcomes of work-family conflict can be physical, such as fatigue, sleep deprivation, or increased susceptibility to infections (Ironson, 1992; Frone, Russell & Barnes, 1996; Frone & Russell, 1995; Frone & Russell, 1997; Goldsmith, 1989), or psychological, such as burnout, stress, and frustration (Wethington & Kessler, 1989; Warp, 1990; Repetti, Matthews & Waldron, 1989; Klitzman, House, Izrael, & Mero, 1990; Barnett, Marshall, Raudenbush & Brennan, 1993; Dekker & Schaufeli, 1995). Researchers now distinguish between work-to-family conflict (WFC when work demands interfere with or take a toll on the family) and family-to-work conflict (FWC when family demands interfere with work) and have paid particular attention to the impact of work-to-family interference (e.g., Frone et al., 1992; Kossek & Ozeki, 1998; Netemeyer et al., 1996; Thomas & Ganster, 1995). Allen, Herst, Bruck, and Sutton (2000) provided a comprehensive review of the consequences associated with work interference with family; they categorized potential outcomes as work-related, non-work-related, and stress-related. In terms of work-related outcomes, most studies find that job satisfaction goes down as WFC goes up (see also review by Kossek & Ozeki, 1998). Intention to turn over was the work-related issue most clearly related to WFC. Relationships have also been reported between WFC and job burnout, work alienation, job tension, and organizational commitment. In terms of non-work-related outcomes, life satisfaction, marital adjustment and satisfaction, and family satisfaction all appear to be negatively affected by work interference with family. WFC was also related to a variety of stress-related outcomes such as general mental health, feeling of self-worth, depression, anxiety and irritability, and life strain. Physical problems associated with WFC include poor appetite, elevated blood pressure, fatigue, nervous tension, and several overall measures of physical health and energy.

Sexual Orientation

Discrimination and harassment based on sexual orientation, although not directly connected to gender, are related to beliefs about gender roles in society. As part of a larger, multi-site, longitudinal health study, Krieger and Sidney (1997) found that among those participants who indicated that they had had sex with a same-sex partner, 33% of the black women, 39% of the black men, 55% of the white women, and 56% of the white men reported experiencing discrimination based on sexual orientation. They found some health-related consequences

Introduction

correlated with this discrimination; however, it was difficult to isolate the effects of sexual orientation-related discrimination since the vast majority of black participants had also experienced racial discrimination and over 80% of the women had also experienced gender discrimination. Research on specifically physical health effects of discrimination based on sexual orientation are somewhat mixed and seem to vary by race and educational status (Huebner, 2002; Krieger & Sidney, 1997); some severe mental health effects have been highly related to experiences of discrimination. Huebner (2002) found that, in a racially diverse sample of 361 gay men, perceived discrimination was associated with depressive symptoms, including suicidal ideation. In a sample of gay and bisexual Latino men, Diaz, Ayala, & Bein (2001) found that discrimination was a strong predictor of psychological symptoms such as suicidal ideation, anxiety, and depression. More specific to the workplace, Waldo (1999) found that people who were "out" at work experienced more discrimination based on sexual orientation and experienced more physical symptoms. Interestingly, he also found that those who chose to hide their sexual orientation experienced what he called "indirect discriminatory events" (e.g., feeling that it is necessary to "act straight") and that experiencing this indirect discrimination was also associated with more symptoms.

Interacting Influences

While this document focuses primarily on issues of race and gender (and to some extent on sexual orientation and work-family issues), it is critical to recognize that these factors interact with one another and are also connected to other dimensions of diversity such as social class and disability. For example, women of color experience the negative effects of both gender- and race-related discrimination in the workplace (Evans & Herr, 1991; Piotrkowski 1998; Xu & Leffler, 1996). Moradi's (2002) study of African American women points to substantial overlap in the impacts of racist and sexist treatment on mental health, and her work supports the notion that these dimensions of discrimination are intertwined and not merely additive.

It is particularly important to acknowledge how race and gender dynamics can overlay social class (Krieger, 2003; Nazroo, 2003). A growing social epidemiology literature addresses the inverse relationship between socioeconomic position and health. The study of social disparities in health calls for the development and application of theoretical frameworks that can support data collection and analysis of the impact of social organization upon population health (Krieger, 1995, 1999; Levins & Lopez 1999).

While there is substantial agreement on the strength and direction of the relationship between socioeconomic position and health outcomes, explanations for this relationship enjoy much less unanimity of opinion. Marmot (1999), for one, has argued for the central role of low control over one's life circumstances, especially in the workplace. Paid employment is a major structural link between education and income: education is a major determinant of people's jobs, which determine their salaries as well as at least some part of the economic assets they accumulate. Employment is also a likely important mediator of socioeconomic disparities in health, because working conditions vary markedly across socioeconomic level (Borg & Kristensen, 2000). Moreover, the workplace is a prime locus for the experience of social status and of discrimination.

Introduction

For example, women and people of color are not typically found in equal proportions across all levels of (job) status within an institution. Thus they may experience adverse situations at work that are actually more a function of their position within the hierarchy than directly a result of being female or minority. Traditional socioeconomic indices of occupation (Marmot 1989), education (Feldman, Makve, Kleinman, 1989) and household income (Duleep, 1986) have all been linked to general health outcomes, with lower-status individuals and families faring worse than those who are relatively advantaged. To the extent that racism and discrimination are also factors in hiring decisions, wage determination, and promotions, research showing the negative health effects of wage discrimination is also relevant here (Darity, 2003). These elements of social disadvantage rooted in race and gender issues are clearly relevant to health. These multiple dimensions undoubtedly interact; in some studies racial differences in physical and mental health are less pronounced when adjusted for income and education, although perceived racial discrimination is still a contributing factor to health status (Williams et al., 1997; Kwate, Valdimarsdottir, Guevarra, & Bovbjerg, 2003).

MEASURES DESCRIBED IN THIS COMPENDIUM

In sum, past research across several disciplines has revealed that gender- and race-related factors such as values, biases, harassment, discrimination, and lack of support for work-family balance can affect physical and mental health. However, these features of the work environment have rarely been included simultaneously with the study of other workplace conditions. Thus, knowledge about correlations among them is still very limited, as is knowledge about potential confounding and interactions.

One barrier to increased inclusion of these dimensions in occupational health research is the limited availability and lack of awareness of appropriate measurement instruments (Meyer, 2003). Much of the research on discrimination has been conducted by investigators in the fields of psychology and sociology, yet lack of communication among disciplines means that occupational health researchers often have little knowledge of relevant instruments developed in other fields. The primary purpose of this document is to consolidate information about relevant survey instruments that assess workplace race and gender dynamics and to bring them to the attention of occupational health scientists.

Another challenge to incorporating diversity issues into occupational health research is that differential treatment manifests itself at multiple levels, as discussed above. Measurement issues and approaches are clearly different for the varied manifestations of bias and discrimination (e.g., individual workers' beliefs versus organizational practices), and thus a wide range of strategies is required to assess relevant dimensions. For example, since workplace conditions and practices are influenced by shared beliefs about who and what is valued by an organization, both workers' attitudes and organizational values related to gender and race/ethnicity can be helpful for capturing relevant diversity dynamics. Additionally, systemic forms of bias (e.g., as evidenced by sexual and racial segregation, different job assignments, differential rates of promotion, and lower organizational responsiveness to complaints) are also important to assess (Browne, 1997; Lott, 1995; Weber & Higginbotham, 1997).

Introduction

This compendium includes measures at multiple levels. However, instruments that assess perceptions and personal descriptions of experiences from individual workers are far more common than more systemic indicators. In reviewing these measures, it is also apparent that samples used to assess psychometric strengths are often not diverse in terms of race, ethnicity, gender or sexual orientation, thus limiting our knowledge of their usefulness with exactly the populations we wish to reach. Thus while a major goal is to emphasize the availability of useful measures, it is also hoped that this collection will demonstrate a need for a wider range and variety of approaches and will stimulate the development of new instruments for assessing employer attitudes and workplace practices and policies.

The criteria for inclusion of measures in this compendium were 1) topical relevance, 2) at least some evidence of psychometric strengths, and 3) use in at least one published study. Our search for candidate measures was done using a snowball approach following up on leads identified through studies on related topics and soliciting suggestions from researchers associated with relevant organizations. We searched the formal social science research literature for articles about measures assessing the following domains: 1) racism and racial/ethnic prejudice, 2) sexism and sexual harassment, 3) gender and racial discrimination, 4) work-family integration and balance, and 5) support for diversity in the workplace/workforce. We reexamined the studies included in our literature review and followed leads to the measures used. We included some measures that we were aware of from our own past research on related topics. In addition, we put out an open call for colleagues to nominate measures by posting the request on lists for the Society for the Psychological Study of Social Issues and the Society for Community Research and Action. We also consulted with the Centers for Disease Control and Prevention's Measure of Racism Working Group to identify any gaps.

We were able to identify 46 measures in the literature that met our criteria. There were two additional dilemmas that shaped some of our choices about which measures to include. First, we found that a very common approach to measuring experiences of discrimination is to ask a straightforward question, essentially, "Did you experience discrimination or not?" In some cases, a single question is used; in others, a few variations are included (e.g., asking "experienced it ever?" then "experienced it in this job?" Participants may also be asked to indicate whether they have been discriminated against in each of several contexts). We have included a few such measures, primarily those that had undergone some psychometric analysis; however, we chose not to include all the variations we saw adapted by individual researchers. Second, many reasonable discrimination and harassment measures have not been developed or ever used in work settings. Some of these are very specific to other settings (most typically academic settings). We did not completely restrict our search or our entries to workplace-specific measures, in part because this would have produced a very short list of instruments. On the contrary, we have included a number of scales that were developed and used in other types of settings but could be adapted for use in occupational health research.

Each entry includes a general description of the measure, sample items, and information about various psychometric strengths and limitations. We first summarized the information we were able to locate in the literature and then sent our draft entries to the scale's authors, requesting their assistance in both checking the entry and providing additional information. We received comments back from about half of the authors. We have included the primary references for

Introduction

each scale and, when available, information about how to obtain it. We have tried to include enough detail to help researchers make informed choices, even though we do not make explicit recommendations for use of one measure over another. Actual copies of measures are not included.

In collecting these measures, it became apparent that there are three main types: 1) ratings of attitudes or beliefs about race, gender, work-life, or sexual orientation (could be general or one's own beliefs or observations) 2) assessments of one's own experiences of bias, harassment, or discrimination (including frequency, severity, and stressfulness), and 3) ratings of the climate or general practices within an organization or group. In this compendium, each measure was assessed by validity and reliability, which are two important standards to consider in constructing and evaluation survey instruments. We assessed three types of validity for each measure: content, construct, and concurrent validity. The next page presents definitions of validity and reliability applied in this document. Following the list of definitions is a section on "Summary of Measures" indicating which of these three types of assessments are incorporated within each measure. We have also noted here which scales include items specifically designed for workplace studies.

Introduction

VALIDITY AND RELIABILITY: DEFINITIONS APPLIED IN THIS DOCUMENT

Content Validity

- The extent to which the scale has appropriate coverage of the subject matter – i.e., does it adequately sample the universe of possible items?
- Includes actions taken to ensure adequate sampling of possible items for the desired content area.
- Face validity and subjective evaluation by expert judges about appropriateness.
- Common approaches include use of focus groups, interviews, or pilot surveys to gather items based on participants' experiences.

Construct Validity

- The extent to which the scale is a good measure of the theoretical constructs that underlie it – i.e., does the scale measure what it says it measures?
- Does it have the relationship to other variables (including demographics) that theories would predict it to have?
- Underlying constructs are often assessed through factor analysis, principal components analysis, etc.

Concurrent Validity

- The relationship between the scale and an external criterion, ideally something that is already accepted as a gold standard for the same phenomenon. Sometimes also referred to as "criterion validity."
- Most often expressed as the correlation between scale scores and scores on a similar already-validated measure of the same phenomenon.

Reliability

- Internal reliability is most commonly expressed as Cronbach's alpha coefficient and sometimes with split-sample reliability
- Test-retest reliability can be assessed using raw percentage of concordant replies or another statistical measure of agreement.

Summary of Measures

Summary of Measures

GENERAL DIVERSITY MEASURES, DIVERSITY CLIMATE, MULTIPLE ISMS

AUTHOR	NAME OF SCALE	CONSTRUCT MEASURED	BRIEF DESCRIPTION # of items (scale anchors); subscales	TYPE OF MEASURE			WORK-RELATED ITEMS	
				Attitudes	Experiences	Climate	Some	Full
Diaz et al. (2001) See page 39	Homophobia and Racism Scales	Experiences of homophobia and racism both as children and as adults	Homophobia: 11 items (never to many times) Racism: 10 items (never to many times)		X			
Godfrey et al. (2000) See page 41	Modified Godfrey-Richman ISM Scale (M-GRISMS-M)	Stereotypes, prejudice and discrimination towards various ethnic, religious groups; sexist and heterosexist attitudes	33 items (yes/no; various Likert rankings) Four subscales: 1. Racism subscale 2. Religion subscale 3. Sexism subscale 4. Heterosexism subscale	X				

LEGEND

TYPE OF MEASURE	Attitudes	Ratings of attitudes or beliefs about race or gender – could be general or own beliefs, observations, etc.
	Experiences	Ratings of own experiences of discrimination.
	Climate	Ratings of the climate or general practices within an organization.
WORK-RELATED ITEMS	Some	There are some items on the scale relevant to work or workplace issues.
	Full	The scale is entirely focused on work-related issues.

Summary of Measures

GENERAL DIVERSITY MEASURES, DIVERSITY CLIMATE, MULTIPLE ISMS

AUTHOR	NAME OF SCALE	CONSTRUCT MEASURED	BRIEF DESCRIPTION # of items (scale anchors); subscales	TYPE OF MEASURE			WORK-RELATED ITEMS	
				Attitudes	Experiences	Climate	Some	Full
Jeanquart-Barone & Sekaran (1996) See page 44	Perceived Supervisory Discrimination	The perceived unfair treatment by immediate supervisor	8 items (agree-disagree) The items can be asked for race or gender discrimination		X			X
Kossek & Zonia (1993) See page 46	Diversity Climate	Perceptions of climate supportive of diversity	16 items (agree to disagree; much higher to much lower; better chance to less chance) Five subscales: 1. Value efforts to promote diversity 2. Attitudes toward qualifications of racioethnic minorities 3. Attitudes toward qualifications of women 4. Equality of department support of racioethnic minorities 5. Equality of department support of women (3 items)			X		X

Summary of Measures

GENERAL DIVERSITY MEASURES, DIVERSITY CLIMATE, MULTIPLE ISMS

AUTHOR	NAME OF SCALE	CONSTRUCT MEASURED	BRIEF DESCRIPTION # of items (scale anchors); subscales	TYPE OF MEASURE			WORK-RELATED ITEMS	
				Attitudes	Experiences	Climate	Some	Full
Larkey (1996) See page 49	Workforce Diversity Questionnaire (WDQ)	Communicative interactions in diverse workgroups	15 items (agree-disagree) Four subscales: 1. Inclusion 2. Ideation 3. Understanding 4. Treatment			X		X
Ponterotto et al. (1995) See page 53	Quick Discrimination Index	Attitudes toward racial diversity and women's equality	30 items (agree-disagree) Three factors: 1. Attitudes about diversity 2. Personal attitudes about racial diversity 3. Gender-based attitudes	X				

Summary of Measures

RACE, RACISM, ETHNICITY, RACIAL DISCRIMINATION & RELATED MEASURES

AUTHOR	NAME OF SCALE	CONSTRUCT MEASURED	BRIEF DESCRIPTION # of items (scale anchors); subscales	TYPE OF MEASURE			WORK-RELATED ITEMS	
				Attitudes	Experiences	Climate	Some	Full
Barbarin & Gilbert (1981) See page 61	Institutional Racism Scale (IRS)	Individual perceptions of self and the organization: how individuals construe institutional racism, engage in anti-racism, and view organizational commitment to the reduction of racism	72 items (various Likert rating systems) Six subscales: Self-attribute subscales: 1. Indices of Racism scale (8 items) 2. Strategies of Reducing Racism-Use scale (7 items) 3. Strategies of Reducing Racism-Effectiveness scale (11 items) 4. Personal Efforts to Reduce Racism scale(20 items) Organizational attribute subscales: 5. Agency Climate scale (6 items) 6. Management/Administrative Efforts to Reduce Racism scale (20 items)	X	X	X	X	

Summary of Measures

RACE, RACISM, ETHNICITY, RACIAL DISCRIMINATION & RELATED MEASURES

AUTHOR	NAME OF SCALE	CONSTRUCT MEASURED	BRIEF DESCRIPTION # of items (scale anchors); subscales	TYPE OF MEASURE			WORK-RELATED ITEMS	
				Attitudes	Experiences	Climate	Some	Full
Cervantes et al. (1991) See page 66	Hispanic Stress Inventory (HSI)	Psychosocial stress among Hispanic adults	<u>Immigrant Version</u> 73 items (yes/no, then not at all stressful to extremely stressful) 5 subscales: 1. Occupational/Economic Stress 2. Parental Stress 3. Marital Stress 4. Immigration Stress 5. Family/Cultural Stress <u>U.S.-Born Version</u> 59 items (yes/no, then not at all stressful to extremely stressful) 4 subscales: 1. Occupational/Economic Stress 2. Parental Stress 3. Marital Stress 4. Family/Cultural Stress		X		X	
Green (1995) See page 72	Perceptions of Racism Scale (PRS)	Perceptions of racism against African Americans	20 items (agree to disagree) Single dimension = observed racism	X	X			

Summary of Measures

RACE, RACISM, ETHNICITY, RACIAL DISCRIMINATION & RELATED MEASURES

AUTHOR	NAME OF SCALE	CONSTRUCT MEASURED	BRIEF DESCRIPTION # of items (scale anchors); subscales	TYPE OF MEASURE			WORK-RELATED ITEMS	
				Attitudes	Experiences	Climate	Some	Full
Harrell (1997) & Harrel et al. (1997) See page 75	The Racism and Life Experience Scales (RaLES)	The RaLES is a comprehensive set of scales that measures racism-related stress, coping, socialization, and attitudes. Only the scales for frequency and stressfulness of racism-related experiences are described here.	Six primary scales and 1 brief scale available: 1. Perceived influence of race 2. Domains of racism experience 3. Daily life experiences 4. Racism experiences 5. Group impact 6. Life experiences and stress 7. Brief scale (9 items) (different rating anchors for different scales)		X		X	
Hughes & Dodge (1997) See page 84	Workplace Racial Bias	Experiences of interpersonal and institutional discrimination at work	13 items (agree to disagree) Two subscales: 1. Institutional racism 2. Interpersonal prejudice		X			X
Krieger (1990) & Krieger & Sidney (1996) See page 87	Krieger Measure of Experiences of Discrimination	Self-reported experiences of and responses to racial discrimination	7 items (yes/no) Experiences of discrimination in variety of settings/situations		X		X	
Landrine & Klonoff (1996) See page 90	Schedule of Racist Events (SRE)	Experience of specific racist events	18 items (scales vary) Specific racist events rated for: frequency in last year; in lifetime (never happened-happens all the time) and, degree of stressfulness (not at all stressful-very stressful)		X		X	

Expanding our Understanding of the Psychosocial Work Environment:
A Compendium of Discrimination, Harassment, and Work-Family Issues

Summary of Measures

RACE, RACISM, ETHNICITY, RACIAL DISCRIMINATION & RELATED MEASURES

AUTHOR	NAME OF SCALE	CONSTRUCT MEASURED	BRIEF DESCRIPTION # of items (scale anchors); subscales	TYPE OF MEASURE			WORK-RELATED ITEMS	
				Attitudes	Experiences	Climate	Some	Full
McConahay (1986) & McConahay et al. (1981) See page 93	Modern Racism Scale	Racial attitudes (based on belief that discrimination no longer exists)	14 items (agree to disagree) Two dimensions 1. Old-fashioned racism 2. Modern racism	X	X		X	
McNeilly et al. (1996) See page 97	Perceived Racism Scale (PRS)	Perceived exposure to racism	51 items total Two approaches to ratings: *Frequency of exposure to racist events in 4 domains (43 items): 1. job 2. academic 3. public 4. racist statements (Not applicable to several times per day) *Emotional appraisal of each event (8 items)		X		X	
Plant & Devine (1998) See page 100	Motivation to respond without Prejudice	Sources of internal and external motivations to respond without prejudice toward blacks	10 items (agree to disagree) Two subscales (5 items each): 1. Internal Motivation to Respond without Prejudice (IMS) 2. External Motivation to Respond without Prejudice (EMS)	X				
Salgado de Snyder (1987) See page 105	Acculturative Stress Scale (ACS)	Acculturative stress in 5 domains: familial, marital, social, financial, and environmental	12 items (not stressful to very stressful)		X			

Summary of Measures

RACE, RACISM, ETHNICITY, RACIAL DISCRIMINATION & RELATED MEASURES

AUTHOR	NAME OF SCALE	CONSTRUCT MEASURED	BRIEF DESCRIPTION # of items (scale anchors); subscales	TYPE OF MEASURE			WORK-RELATED ITEMS	
				Attitudes	Experiences	Climate	Some	Full
Terrell & Terrell (1981) See page 108	Cultural Mistrust Inventory (CMI)	Beliefs about the extent to which African Americans should trust Euro Americans/whites	48 items (agree to disagree) Four subscales: 1. Political/legal system 2. Work interactions 3. Education and training 4. Interpersonal/social	X			X	
Thompson, et al. (1990) See page 111	Racism Reaction Scale (RRS)	Sense of being personally threatened, differentially treated, or singled out for differential treatment	6 items (agree to disagree)		X			
Utsey & Ponterotto (1996) See page 113	Index Of Race-Related Stress (IRRS)	The occurrence of and stress associated with specific events of racism and discrimination	46 items (never happened to happened and I was extremely upset) Four subscales: 1. Cultural racism 2. Institutional racism 3. Individual racism 4. Collective racism		X		X	
Williams et al. (1997) See page 118	Race-Related Stress	Experiences of lifetime discrimination and everyday discrimination	12 items (scales vary) Two subscales: 1. Lifetime discrimination (3 items) (none-three or more events) 2. Everyday discrimination (9 items) (never-often)		X		X	

Summary of Measures

SEXISM & SEX DISCRIMINATION

AUTHOR	NAME OF SCALE	CONSTRUCT MEASURED	BRIEF DESCRIPTION # of items (scale anchors); subscales	TYPE OF MEASURE			WORK-RELATED ITEMS	
				Attitudes	Experiences	Climate	Some	Full
Benson & Vincent (1980) See page 123	Sexist Attitudes Toward Women Scale (SATWS)	Tendency toward and support for sexist attitudes	40 items (agree to disagree)	X				
Glick & Fiske (1996) See page 127	The Ambivalent Sexism Inventory (ASI)	Hostile and benevolent sexism toward women	22 items (agree to disagree) Two subscales: 1. Hostile sexism 2. Benevolent sexism	X				
Glick & Fiske (1999) See page 132	The Ambivalence Toward Men Inventory (AMI)	Women's hostile and benevolent prejudices toward men	20 items (agree-disagree) Two subscales: 1. Hostility toward men 2. Benevolence toward men	X				
Klonoff & Landrine (1995) See page 135	Schedule of Sexist Events (SSE)	Lifetime and recent sexist discrimination in women's lives	20 items (scales vary) Specific sexist events rated for frequency in last year, in lifetime (never happened to happens all the time) and degree of stressfulness (not at all stressful to very stressful)		X		X	
Pinel (1999) See page 139	Stigma Consciousness Questionnaire (SCQ)	The extent to which people focus on their stereotyped status	10 items (agree to disagree)	X	X			

Summary of Measures

SEXISM & SEX DISCRIMINATION

AUTHOR	NAME OF SCALE	CONSTRUCT MEASURED	BRIEF DESCRIPTION # of items (scale anchors); subscales	TYPE OF MEASURE			WORK-RELATED ITEMS	
				Attitudes	Experiences	Climate	Some	Full
Riger, Stokes, Raja & Sullivan (1997) See page 146	Working Environment For Women In Academic Settings	Perceptions of attitudes toward women faculty in university settings	35 items (agree to disagree; not at all likely to very likely) Three subscales: 1. Differential treatment 2. Balancing work and personal obligations 3. Sexist attitudes and comments		X			X
Stokes, Riger, & Sullivan (1995) See page 149	Working Environment For Women In Corporate Settings	Perceptions of attitudes toward women in the work environment	36 items (agree to disagree; not at all likely to very likely) Five subscales: 1. Dual standards and opportunities (10 items) 2. Sexist attitudes and comments (7 items) 3. Informal socializing (7 items) 4. Balancing work and personal obligations (9 items) 5. Remediation policies and practices (3 items)		X			X

Summary of Measures

SEXISM & SEX DISCRIMINATION

AUTHOR	NAME OF SCALE	CONSTRUCT MEASURED	BRIEF DESCRIPTION # of items (scale anchors); subscales	TYPE OF MEASURE			WORK-RELATED ITEMS	
				Attitudes	Experiences	Climate	Some	Full
Swim et al. (1995) See page 153	Old-Fashioned And Modern Sexism Scale	Old-fashioned sexism and modern sexism	13 items (agree to disagree) Two subscales: 1. Old-Fashioned Sexism (5 items) 2. Modern Sexism (8 items)	X			X	
Swim et al. (2001) See page 156	Everyday Sexism	Observations of daily instances of sexism experienced both by self and by others	Diary approach		X		X	
Tougas et al. (1995) See page 159	General Attitudes Toward Affirmative Action (AA) And Men's Collective Interest (CI)	Attitudes toward affirmative action	3 items (AA) (agree to disagree) 6 items (CI) (agree to disagree)	X				X
Tougas et al. (1995) See page 162	Neosexism Scale	Conflict between egalitarian values and residual negative feeling toward women	11 items (agree to disagree)	X			X	

Summary of Measures

SEXUAL HARASSMENT

AUTHOR	NAME OF SCALE	CONSTRUCT MEASURED	BRIEF DESCRIPTION # of items (scale anchors); subscales	TYPE OF MEASURE				WORK-RELATED ITEMS	
				Attitudes	Experiences	Climate		Some	Full
Bingham & Scherer (1993) See page 167	Responses to Sexual Harassment and Satisfaction With The Outcome	Three constructs: Work climate regarding sexual harassment; Responses to sexual harassment; Satisfaction with the outcome	<u>Work Climate</u> 3 items (strongly agree to strongly disagree) <u>Responses to sexual harassment</u> 1. General (5-item checklist) 2. Communication strategies (checklist) <u>Satisfaction with outcome</u> 1 item (definitely not satisfied, yes, definitely satisfied)		X	X			X
Cortina (2001) See page 170	Sexual Experiences Questionnaire – Latinas (SEQ-L)	Experiences of sexual harassment among Latinas	20 items (never to most of the time) Three subscales: 1. Sexist hostility 2. Sexual hostility 3. Unwanted sexual attention		X			X	

Summary of Measures

SEXUAL HARASSMENT

AUTHOR	NAME OF SCALE	CONSTRUCT MEASURED	BRIEF DESCRIPTION # of items (scale anchors); subscales	TYPE OF MEASURE			WORK-RELATED ITEMS	
				Attitudes	Experiences	Climate	Some	Full
Fitzgerald, Gelfand, & Drasgow (1995) See page 173	Sexual Experience Questionnaire (SEQ-W)	Experience of sexual harassment in the workplace	20 items (never to many times) Three subscales: 1. Gender harassment 2. Unwanted sexual attention 3. Sexual coercion		X			X
Hulin, Fitzgerald, & Drasgow (1996) See page 177	Organizational Tolerance For Sexual Harassment Inventory (OTSHI)	Perceptions of likelihood of organizational reactions to various forms of harassment	18 items Three subscales: 1. Risk of reporting 2. Likelihood of being taken seriously 3. Probability of sanctions			X		X
Murdoch & McGovern (1998) See page 180	Sexual Harassment Inventory (SHI)	Experiences of sexual harassment	20 items (yes/no) and one open-ended question Responses can be severity weighted Three subscales: 1. Hostile environment 2. Quid pro quo 3. Criminal sexual misconduct		X	X		X

Summary of Measures

WORK-FAMILY/WORK-LIFE RELATED MEASURES

AUTHOR	NAME OF SCALE	CONSTRUCT MEASURED	BRIEF DESCRIPTION # of items (scale anchors); subscales	TYPE OF MEASURE			WORK-RELATED ITEMS	
				Attitudes	Experiences	Climate	Some	Full
Bacharach, Bamberger, & Conley (1991) See page 184	Work-Family Conflict	Interrole conflict where role pressures from work and family (home) domains are mutually incompatible in some respect	4 items (seldom/never to almost always)		X			X
Barnett & Gareis (under review) See page 189	Parental After-School Stress (PASS)	Degree to which employed parents are concerned about the welfare of their school-aged children during the after-school hours	10 items (not at all to extremely)		X			

Summary of Measures

WORK-FAMILY/WORK-LIFE RELATED MEASURES

AUTHOR	NAME OF SCALE	CONSTRUCT MEASURED	BRIEF DESCRIPTION # of items (scale anchors); subscales	TYPE OF MEASURE			WORK-RELATED ITEMS	
				Attitudes	Experiences	Climate	Some	Full
Barnett et al. (1999) See page 192	Work Schedule Fit	Degree to which work schedule meets own and family needs	11 items (extremely poorly to extremely well) Three domains: 1. Fit of own schedule for oneself (self/self schedule fit) 2. Fit of own schedule for other family members (self/family schedule fit) 3. Fit of partner's schedule, if applicable, for all family members (partner/family schedule fit)		X			X
Behson (2002) See page 195	Informal Work Accommodations To Family (IWAF)	Ways in which employees temporarily and informally adjust their usual work patterns in an attempt to balance their work and family responsibilities	16 IWAF behaviors (never to very often)		X			X

Summary of Measures

WORK-FAMILY/WORK-LIFE RELATED MEASURES

AUTHOR	NAME OF SCALE	CONSTRUCT MEASURED	BRIEF DESCRIPTION # of items (scale anchors); subscales	TYPE OF MEASURE			WORK-RELATED ITEMS	
				Attitudes	Experiences	Climate	Some	Full
Bohen & Viveros-Long (1981) See page 199	Job-Family Role Strain Scale	Perceptions of stress related to job and family obligations	19 items (always to never) Five dimensions: 1. Ambiguity about norms (3 items) 2. Socially structured insufficiency of resources for role fulfillment (3 items) 3. Low rewards for role conformity (3 items) 4. Conflict about normative phenomena (4 items) 5. Overload of role obligations (6 items)		X			X
Bohen & Viveros-Long (1981) See page 203	Family Management Scale	Difficulty in managing the logistics of family life	21 items (level of difficulty)		X			
Cedillo-Becerril (1999) See page 207	Spillover Between Home And Job Responsibilities	Lack of balance between job and family responsibilities	2 items (agree to disagree)		X			X

Summary of Measures

WORK-FAMILY/WORK-LIFE RELATED MEASURES

AUTHOR	NAME OF SCALE	CONSTRUCT MEASURED	BRIEF DESCRIPTION # of items (scale anchors); subscales	TYPE OF MEASURE			WORK-RELATED ITEMS	
				Attitudes	Experiences	Climate	Some	Full
Eaton (1999) See page 209	Work-Family Policies—Perceived Management Support and Usability	The extent to which the organization supports employee efforts to balance work and family	Two scales 1. General perceptions of management support for policies (7 items; not at all to a great deal) 2. Assessment of availability of specific policies (10 items; yes/no on various dimensions)			X		X
Friedman & Greenhaus (2000) See page 212	Employer Support For Family	Organizational support for work and family balance	5 items (agree to disagree)			X		X
Friedman & Greenhaus (2000) See page 214	Work-Family Interference And Tradeoffs	The perception that the demands of the work role and the demands of the family role interfere with one another	13 items (agree to disagree) Interference: three subscales: 1. Behavioral work interference with family (2 items) 2. Work interference with family (4 items) 3. Family interference with work (5 items) Tradeoffs (2 items)		X			X

Summary of Measures

WORK-FAMILY/WORK-LIFE RELATED MEASURES

AUTHOR	NAME OF SCALE	CONSTRUCT MEASURED	BRIEF DESCRIPTION # of items (scale anchors); subscales	TYPE OF MEASURE				WORK-RELATED ITEMS	
				Attitudes	Experiences	Climate		Some	Full
Friedman & Greenhaus (2000) See page 217	Adjustment Of Work Schedule	Adjustment of work schedule for family and personal reasons	4 items (never to frequently)		X				X
Frone & Yardley (1996) See page 219	Work-Family Conflict	Interference of job with the family life and of the family with the job	12 items Two subscales: 1. Interference of job with family life (6 items) 2. Interference of family with job (6 items)		X				X

Expanding our Understanding of the Psychosocial Work Environment:
A Compendium of Discrimination, Harassment, and Work-Family Issues

Summary of Measures

WORK-FAMILY/WORK-LIFE RELATED MEASURES

AUTHOR	NAME OF SCALE	CONSTRUCT MEASURED	BRIEF DESCRIPTION # of items (scale anchors); subscales	TYPE OF MEASURE			WORK-RELATED ITEMS	
				Attitudes	Experiences	Climate	Some	Full
Geurts et al. (in preparation) Wagena & Geurts (2000) See page 222	Survey Work-Home Interaction-Nijmegen (SWING)	The extent to which one's functioning in one domain is influenced by demands from the other domain	27 items (almost never to almost always) Four types of work-home interaction (WHI): 1. Work negatively influencing home (WHI-) 2. Home negatively influencing work (HWI-) 3. Work positively influencing home (WHI+) 4. Home positively influencing work (HWI+)		X			X
Gutek, Searle, & Klepa (1991) See page 227	Work-Family Conflict	Extent to which work demands interfere with family and family demands interfere with work	8 items (agree to disagree) Two subscales: 1. Work interference with family (4 items) 2. Family interference with work (4 items)		X			X

Summary of Measures

Work-Family/Work-life Related Measures

AUTHOR	NAME OF SCALE	CONSTRUCT MEASURED	BRIEF DESCRIPTION # of items (scale anchors); subscales	TYPE OF MEASURE				WORK-RELATED ITEMS	
				Attitudes	Experiences	Climate		Some	Full
Hughes & Galinski (1994) See page 230	Work Family Interference	Work-family interference	14 items (never to very often) Two subscales: 1. Family role difficulty (8 items) 2. Job role difficulty (6 items)		X				X
Kopelman, Greenhaus, & Connoly (1983) See page 233	Interrole Conflict Scale	Conflict between work and family roles	8 items (agree to disagree)		X				X
Netemeyer, Boles, & McMurrian (1996) See page 237	Work-Family Conflict And Family-Work Conflict Scales	Conflict generated in family life because of work, and conflict generated at work because of family	10 items (agree to disagree) Two subscales: 1. Work to family conflict (5 items) 2. Family to work conflict (5 items)		X				X

Summary of Measures

WORK-FAMILY/WORK-LIFE RELATED MEASURES

AUTHOR	NAME OF SCALE	CONSTRUCT MEASURED	BRIEF DESCRIPTION # of items (scale anchors); subscales	TYPE OF MEASURE				WORK-RELATED ITEMS	
				Attitudes	Experiences	Climate		Some	Full
Small & Riley (1990) See page 241	Worker Perception Of Work Spillover	Spillover of work into home/personal life	20 items (agree to disagree) Four subscales: 1. Spillover into the marital relationship 2. Spillover into the parent-child relationship 3. Spillover into leisure time 4. Spillover into household tasks		X				X
Stephens, & Sommer (1996) See page 244	Work-To-Family Conflict	Extent to which work demands affect family	14 items (agree to disagree) Three subscales: 1. Time-based conflict (4 items) 2. Strain-based conflict (4 items) 3. Behavior-based conflict (6 items)		X				X

Summary of Measures

Sexual Orientation/Homophobia

Author	Name of Scale	Construct Measured	Brief Description # of items (scale anchors); subscales	Type of Measure				Work-Related Items	
				Attitudes	Experiences	Climate		Some	Full
Krieger & Sidney (1997) See page 249	Discrimination Based On Sexual Orientation Questionnaire	Self-reported experiences of discrimination based on sexual orientation	7 items (yes/no) Experiences of discrimination in a variety of settings/situations		X			X	
Larsen, Reed, & Hoffman (1980) See page 251	Heterosexual Attitudes Toward Homosexuals (HATH)	Heterosexual attitudes toward homosexuals	20 items (agree to disagree)	X					
Waldo (1999) See page 254	Workplace Heterosexist Experiences Questionnaire (WHEQ)	Employees' experiences of sexual orientation-based harassment and discrimination	22 items (never to most of the time)		X	X			X

General Diversity Measures, Diversity Climate, Multiple Isms

General Diversity Measures, Diversity Climate, Multiple Isms

TITLE OF MEASURE	HOMOPHOBIA AND RACISM SCALES
Source/Primary reference	Diaz, R. M., Ayala, G., Bein, E., Henne, J., & Marin, B. (2001). The impact of homophobia, poverty, and racism on the mental health of gay and bisexual Latino men: Findings from 3 US cities. *American Journal of Public Health, 91*(6), 927-932.
Construct measured	Experiences of homophobia and racism both as children and as adults
Brief description	The Homophobia scale included 11 items and the Racism scale 10 items. Both scales were rated on a 4-point *never* to *many times* scale.
Sample items	Homophobia: • As you were growing up, how often did you feel that your homosexuality hurt and embarrassed your family? • As an adult, how often have you had to pretend that you are straight to be accepted? Racism: • How often have you been turned down for a job because of your race or ethnicity? • In sexual relationships, how often do you find that men pay more attention to your race or ethnicity than to who you are as a person?
Appropriate for whom (i.e. which population/s)	Non-majority, non-heterosexual adults
Translations & cultural adaptations available	English and Spanish versions available
How developed	Qualitative studies preceded the quantitative survey. Approximately 300 gay and bisexual Latino men were interviewed, in a total of 26 focus group discussions, in three cities. The focus group transcripts were used to develop the items for the quantitative survey. The items were refined through pilot testing.
Psychometric properties	*STUDY SAMPLE*

Participants	Demographics	
Sample Size	*n* = 912	
Description	Latino, non-heterosexual men entering social venues in the cities of New York (n = 309), Miami (n = 302), and Los Angeles (n = 301)	
Age	Mean	31.2
Education	Some college or more	64.2%
HIV Status	HIV-positive	21.8%
	HIV-negative	67.3%
	Do not know	10.9%

General Diversity Measures, Diversity Climate, Multiple Isms

TITLE OF MEASURE — HOMOPHOBIA AND RACISM SCALES

RELIABILITY

Internal Consistency

Cronbach's α coefficient for Homophobia and Racism scales.

Scale	α =
Homophobia	.75
Racism	.82

Comments	- Other than face validity, there is minimal information about concurrent or construct validity. - Participants were patrons of Latino gay venues – findings may not apply to men who do not attend gay venues or to men who prefer to attend mainstream gay venues. - Participants were mostly immigrants; the findings may not apply to the experience of U.S.-born Latinos. - Survey data were solely based on self-reports. Thus self-report biases are possible, including the tendency to underreport stigmatized behavior.
Bibliography (studies that have used the measure)	Diaz, R. M., Ayala, G., & Bein, E. (2004). Sexual risk as an outcome of social oppression: Data from a probability sample of Latino gay men in three US cities. *Cultural Diversity and Ethnic Minority Psychology, 10*(3), 255-267. Zea, M. C., Reisen, C. A., Poppen, P. J., & Diaz, R. M. (2003). Asking and telling: Communication about HIV status among Latino HIV-positive gay men. *AIDS and Behavior, 7*(2), 143-152.
Contact Information	Rafael M. Diaz Center for Community Research San Francisco State University 3004 16th St., Suite 301 San Francisco, CA 94103, USA e-mail: rmdiaz@sfsu.edu

General Diversity Measures, Diversity Climate, Multiple Isms

TITLE OF MEASURE	**MODIFIED GODFREY-RICHMAN ISM SCALE (M-GRISMS-M)**
Source/Primary reference	Godfrey, S., Richman, C., & Withers, T. (2000). Reliability and validity of a new scale to measure prejudice: The GRISMS. *Current Psychology, 19*(1), 8-13.
Construct measured	Stereotypes, prejudice and discrimination towards various ethnic and religious groups, as well as sexist and heterosexist attitudes
Brief description	This revised scale includes 33 items. Response options include yes/no, standard Likert ratings, and rankings. It consists of four subscales: 1. Racism subscale (attitudes toward African Americans, Latinos/ Hispanics, Asian Americans, Native Americans, European Americans, as well as general racism) 2. Religion subscale (attitudes towards Christian, Jewish, Moslem, agnostic/atheist persons, as well as general religion questions) 3. Sexism subscale (attitudes toward males and females) 4. Heterosexism subscale (attitudes toward gay men and lesbians as well as general heterosexism)
Sample items	- Native American men are more aggressive and brutal than other men. - Christians are intolerant of people with other religious beliefs. - Sexism was created by women as an excuse for their lower level of success in the business world. - Heterosexual men have a strong desire to dominate and take advantage of women. - Homosexuals should be permitted to teach children in schools.
Appropriate for whom (i.e. which population/s)	Adults
Translations & cultural adaptations available	None known
How developed	An original GRISMS was developed with 90 items (1995, unpublished). It had high reliability and subscale concurrent validity in comparison with other measures of racism, sexism, and heterosexism (Pearson r's ranged from .65 to .76). However, it was very long and time-consuming to complete. Thus the authors worked to develop a 50-item version, called the M-GRISMS. During the study described below, a new revised version (M-GRISMS-M) was developed by eliminating additional items to optimize the internal reliability of each subscale.

General Diversity Measures, Diversity Climate, Multiple Isms

TITLE OF MEASURE   MODIFIED GODFREY-RICHMAN ISM SCALE (M-GRISMS-M)

Psychometric properties

STUDY SAMPLE

Participants	Demographics	
Sample Size	$n = 131$	
Description	Introductory Psychology Students	
Age Range	18 to 23	
Gender	Female Male	$n = 71$ $n = 60$
Race/Ethnicity	European-American	93%
	African American	5%
	Asian or Native American	2%

VALIDITY

Concurrent Validity

The M-GRISMS was compared to the Modern and Old Fashioned Racism Scale (McConahay, Hardee, & Batts, 1981), the Attitudes Toward Women Scale (AWS, Spence, Helmrich, & Strapp, 1973), and the combined Heterosexual Attitudes Toward Homosexuality (HATH Scale, Larsen, Reed & Hoffman, 1980) and Index of Homophobia (IHP, Hudson & Ricketts, 1980) scales.

M-GRISMS Subscale	Modern Racism $r =$	AWS $r =$	HATH/IHP $r =$
Racism	.60 ***		
Sexism		.41 ***	
Heterosexism			.76***

***$p < .001$

M-GRISMS-M Subscale	Modern Racism $r =$	AWS $r =$	HATH/IHP $r =$
Racism Subscale	.75***		
Sexism Subscale		.55 ***	
Heterosexism Subscale			.77***

***$p < .001$

General Diversity Measures, Diversity Climate, Multiple Isms

TITLE OF MEASURE   MODIFIED GODFREY-RICHMAN ISM SCALE (M-GRISMS-M)

RELIABILITY

Internal Consistency

Scale	M-GRISMS $\alpha =$	M-GRISMS - M $\alpha =$
Full Scale	.77	.89
Racism Subscale	.52	.64
Religion Subscale	.17	.40
Sexism Subscale	.44	.52
Heterosexism Subscale	.72	.82

Test-retest Reliability

Scale	M-GRISMS $r =$	M-GRISMS-M $r =$
Full Scale	.66	.89
Racism Subscale	.58	.80
Religion Subscale	.34	.75
Sexism Subscale	.37	.77
Heterosexism Subscale	.66	.81

Comments

- Although the authors have worked to shorten their scale, it remains long, requiring college students about 30 minutes to complete.

- It was developed and tested with college students, so its generalizability to working populations is unknown. However, on the surface, the items would seem transferable.

- It was also developed with a predominantly Euro-American sample, thus usefulness with other groups needs further assessment.

- Internal reliability of the Religion Subscale is quite low and that of the racism and sexism subscales is somewhat marginal.

Bibliography (studies that have used the measure)

Contact Information

Charles L. Richman
Department of Psychology
Wake Forest University
P.O. Box 7778 Reynolds Station
Winston-Salem, NC 27109, USA
Tel: (336) 758-6134

e-mail: richman@wfu.edu

General Diversity Measures, Diversity Climate, Multiple Isms

TITLE OF MEASURE	PERCEIVED SUPERVISORY DISCRIMINATION
Source/Primary reference	Jeanquart-Barone, S., & Sekaran, U. (1996) Institutional racism: An empirical study. *Journal of Social Psychology, 136*(4), 477-482.
Construct measured	Supervisor discrimination and perceived unfair treatment
Brief description	The scale includes 8 items that describe ways discrimination may manifest. Items are rated on a 5-point Likert-type scale from 1 = strongly disagree to 5 = strongly agree. The questions can be asked both as they relate to race discrimination and as they relate to gender discrimination.
Sample items	I believe my RACE/GENDER has had an influence on: <ul><li>my performance evaluations (being judged more critically than others).</li><li>the number of (increased) responsibilities assigned to me.</li><li>the types of jobs given to me (e.g. harder, dirtier work).</li><li>the way I am treated in general.</li></ul>
Appropriate for whom (i.e. which population/s)	Non-majority adult workers
Translations & cultural adaptations available	None known
How developed	The survey items were based on the seven ways that discrimination may manifest itself that have been identified by the Institute of Social Research (ISR).
Psychometric properties	*See study sample below*

STUDY SAMPLE

Participants		Demographics
Sample Size		n = 173
Description		Members of a national minority organization
Race/Ethnicity	Blacks	n = 146
	Others (Asian, Hispanic, American Indian)	n = 30
Gender	*Female*	40%
	Male	60%
Education	*College*	30%
	Some College	45%
Occupation	*Managers*	9%
	Clerical workers	26%
	Others (consultants, technicians, superintendents, nurses)	65%

General Diversity Measures, Diversity Climate, Multiple Isms

TITLE OF MEASURE — PERCEIVED SUPERVISORY DISCRIMINATION

<u>VALIDITY</u>

Concurrent Validity

The results of a regression analysis indicate that perceived supervisory discrimination (intervening variable) had a significant path to institutional racism (dependent variable) as measured by Barbarin and Gilbert's (1981) scale. The path coefficient was .404 ($p < .000001$). Higher levels of perceived supervisory discrimination were associated with respondents' perceptions of higher levels of institutional racism.

<u>RELIABILITY</u>

Internal Consistency

Comments	▪ The instrument was specifically designed to assess the work environment. ▪ The items can be altered for race or gender discrimination. ▪ The research included only a small sample of African-Americans. Only 12% responded to the survey, thus the possibility of non-response bias cannot be ruled out. ▪ Survey data were based on self-reports, thus self-report biases are possible.
Bibliography (studies that have used the measure)	Jeanquart, S. (1991). *Felt conflict of subordinates in vertical dyadic relationships when supervisors and subordinates vary in gender or race.* Doctoral Dissertation. Southern Illinois University at Carbondale. Jeanquart-Barone, S. (1996). Examination of supervisory satisfaction in traditional and nontraditional gender-based reporting relationships. *Sex Roles, 34*(9/10), 717-728.
Contact Information	Sandy Jeanquart Miles, SPHR Professor, Management and Marketing Department College of Business and Public Affairs Business Building South, 413E Murray State University Murray, KY 42071, USA Tel: (270) 762-3401 Fax: (270) 762-3740 e-mail: sandy.miles@murraystate.edu

General Diversity Measures, Diversity Climate, Multiple Isms

TITLE OF MEASURE	DIVERSITY CLIMATE
Source/Primary reference	Kossek, E. E., & Zonia, S. C. (1993). Assessing diversity climate: A field study of reactions to employer efforts to promote diversity. *Journal of Organizational Behavior, 14,* 61-81.
Construct measured	Aspect of the work climate that are supportive of diversity
Brief description	The scale includes 16 items and has 5 subscales: 1. Value efforts to promote diversity (6 items) (5 = strongly agree, 1 = strongly disagree) 2. Attitudes toward qualifications of racioethnic minorities (2 items) (5 = much higher, 1 = much lower) 3. Attitudes toward qualifications of women (2 items) (5 = much higher, 1 = much lower) 4. Equality of department support of racioethnic minorities (3 items) (3 = better chance, 2 = same chance, 1 = less chance) 5. Equality of department support of women (3 items) (3 = better chance, 2 = same chance, 1 = less chance)
Sample items	Value efforts to promote diversity<ul><li>If organization X is to remain an excellent institution it must recruit more minority faculty.</li></ul>Attitudes toward qualifications of racioethnic minorities<ul><li>The scholarly qualifications of minority faculty compared to non-minority faculty in my school/department are _____.</li></ul>Attitudes toward qualifications of women<ul><li>Research productivity of women faculty compared to men faculty in my school/department is _____</li></ul>Equality of department support of racioethnic minorities<ul><li>Compared to non-minority faculty, minority faculty have _____ of having graduate students to assist them.</li></ul>Equality of department support of women<ul><li>Compared to faculty men, faculty women have _____ of getting release from teaching.</li></ul>
Appropriate for whom (i.e. which population/s)	University administrators, faculty and other staff (could be adapted for other types of work settings)
Translations & cultural adaptations available	None known

General Diversity Measures, Diversity Climate, Multiple Isms

TITLE OF MEASURE   DIVERSITY CLIMATE

How developed

Items to assess "diversity climate" were developed by two white female faculty members based on a review of the literature, where 1) climate was conceived as the influence of work contexts on employee behavior and attitudes, which are grounded in perceptions; and 2) it was assumed that people attach meaning to or make sense of clusters of psychologically related events.

The survey was submitted for review to a group of senior administrators, who had requested the study. The administrators included white, black, and Hispanic men and white women of various academic ranks.

Psychometric properties

STUDY SAMPLE

The survey was mailed to all of the office addresses of white women and racioethnic minorities, as well as a random sample of white men with faculty and academic staff status and at least one year's seniority ($n = 1529$). A total of 775 (51%) usable questionnaires were returned.

Population Group	Total Population	Number Sampled	Number Returned	% Returned
Racioethnic Minority Women	87	87	40	46
White Women	629	629	318	51
Racioethnic Minority Men	191	191	83	43
White Men	1,842	600	281	47
Identification Deleted by Respondent			53	
Total	2,749	1,507	775	51

VALIDITY

Construct Validity

Exploratory factor analyses were conducted on 20 items pertaining to diversity. Four distinct factors accounted for most of the variance (66%) among the items. An item was included in a scale if its factor loading exceeded 0.4 and the loading for that item was larger than the loading on any other factor by 0.2. For conceptual reasons, the authors divided a factor related to departmental support into two – one for support of women and one for support of racioethnic minorities. The result was the 5 subscales outlined below.

In multivariate analysis of variance, all five factors were significantly associated with racioethnicity and four with gender (all except attitudes toward qualifications of racioethnic minorities).

General Diversity Measures, Diversity Climate, Multiple Isms

TITLE OF MEASURE | DIVERSITY CLIMATE

RELIABILITY

Internal Consistency

Subscale	Cronbach's α
Value efforts to promote diversity	.77
Attitudes toward qualifications of racioethnic minorities	.71
Attitudes toward qualifications of women	.72
Equality of department support of racioethnic minorities	.74
Equality of department support of women	.90

Comments
- Little empirical study has been conducted on the issue of diversity climates, and this appears to be one of very few scales that tries to directly assess diversity *climate* for workers within an organization.

- While developed for employees within an academic environment, the scale could be adapted for use in other types of organizations.

Bibliography (studies that have used the measure)

Contact Information

Ellen Ernst Kossek
School of Labor & Industrial Relations
Michigan State University
East Lansing, MI 48824, USA

General Diversity Measures, Diversity Climate, Multiple Isms

TITLE OF MEASURE	**WORKFORCE DIVERSITY QUESTIONNAIRE (WDQ)**
Source/Primary reference	Larkey, L. K. (1996). The development and validation of the Workforce Diversity Questionnaire: An instrument to assess interactions in diverse workgroups. *Management Communication Quarterly, 9*(3), 296-338.
Construct measured	Interactions in diverse workgroups
Brief description	The Workforce Diversity Questionnaire (WDQ) consists of 15 items in four subscales that reflect the dimensions of expected behavioral responses to perceived cultural diversity: 1. inclusion (4 items) 2. ideation (4 items) 3. understanding (3 items) 4. treatment (4 items) Items appear to be agree-disagree statements but the scale is not stated explicitly.
Sample items	Inclusion/Exclusion - If someone who is not included in the mainstream tries to get information or makes a request, others stall or avoid helping them out in subtle ways. - It seems that the real reason people are denied promotions or raises is that they are seen as not fitting in. Varied/Conforming Ideation - When people from different backgrounds work together in groups, some people feel slighted because their ideas are not acknowledged. - People are reluctant to get involved in a project that requires them to balance ideas from different gender and racial points of view. Understanding/Misunderstanding - When people who are culturally different or of different genders work together in our group, there is always some amount of miscommunication. - Women and people of color are interpreted differently than white males, even when they say the same thing. Positive/Negative Treatment - Some people in our group are "talked down to" because they are different. - People's different ways of talking or acting cause them to be treated as less competent or smart.

General Diversity Measures, Diversity Climate, Multiple Isms

TITLE OF MEASURE	**WORKFORCE DIVERSITY QUESTIONNAIRE (WDQ)**
Appropriate for whom (i.e. which population/s)	Working adults (men and women of all ethnic groups, by design)
Translations & cultural adaptations available	None known

How developed	Dimensions were derived from existing theoretical literature. Key statements within each dimension were identified through interviews with 15 employees of a high-tech manufacturing firm and 20 from a consumer products manufacturer/distributor. Of the 35 volunteers, 18 were female; 18 were from managerial positions; 16 (46%) were Caucasian, 10 (29%) Hispanic, 7 (20%) African American, and 2 (6%) Asian American.
	
	Questions were designed to generate both positive and negative experiences to gain insight into both poles of each theoretically proposed dimension. Two people coded the interview descriptions of interactions into five theoretically derived dimensions and listed separately any other comments. Inter-rater agreement was 64% in the first set of interviews and 84% in the second set. Six interaction dimensions (including "required work" to describe mundane requirements of the job) were finalized in four context categories. Interviews were repeated until no new statements were generated. In all, 56 items were written, comprising:
	
	    13 for inclusion/exclusion,
	
	    14 for convergence/divergence (later dropped),
	
	    11 for varied/conforming ideation,
	
	    7 for understanding/misunderstanding, and
	
	    11 for positive/negative evaluation (later termed "treatment").
	
	Based on the results of surveys distributed to a snowball sample of students, further wording changes were made to a subset of items, primarily to reflect group observations rather than personal experiences. The resulting instrument was pilot-tested with employees of a high-tech consumer product manufacturer, a social service agency, and a hospital (Pilot Study, see below); further revised to resolve consistency and parallelism issues; and re-evaluated among undergraduate college students (Validation Study).

General Diversity Measures, Diversity Climate, Multiple Isms

TITLE OF MEASURE   WORKFORCE DIVERSITY QUESTIONNAIRE (WDQ)

Psychometric properties

STUDY SAMPLES

Participants	Pilot Study	Validation Study	Participants
Sample Size	n = 280	n = 182	Sample Size
Age	Mean (range)	39 (20-70)	31 (15-62)
Gender	Female/Male	45% / 55%	49% / 50%
Race/Ethnicity	Caucasian	57%	70%
	Hispanic	23%	22%
	African American	10%	2%
	Native American	4%	0%
	Asian American	0%	6%
Position	Hourly wage	35%	
	Salaried	62%	

VALIDITY

Construct Validity

Pilot Study: In factor analysis, the four scales retained were represented by items with strong factor loadings and Cronbach's alpha coefficients between .69 and .80.

Validation Study: The scales were correlated with each other; however, confirmatory factor analysis on all items of the four factors showed that the items did not represent the same factor. The authors also considered that the scales could be ordered by underlying "diversity climate" processes such as dominating attitudes, organizational culture patterns, or situational factors governing the perception of and reactions to those who are different. In a structural equation model, each of the four WDQ factors was indeed found to be influenced by the underlying factor of diversity climate.

Concurrent Validity

Pilot Study: Inclusion, ideation, understanding, and treatment were significantly and positively correlated with 2 "outside" scales: job load (r = .26, .28, .26, and 29, respectively) and color-blindness (r = .35, .46, .44, and .46, respectively).

Validation Study: Of the four "outside" scales included in the instrument, the two scales expected to correlate were strongly associated (power was correlated negatively and cohesion positively with all of the scales), whereas the two scales predicted not to correlate (detail and values) showed only small, mostly nonsignificant correlations.

General Diversity Measures, Diversity Climate, Multiple Isms

TITLE OF MEASURE | WORKFORCE DIVERSITY QUESTIONNAIRE (WDQ)

<u>RELIABILITY</u>

Internal reliability

Validation Study: Cronbach's α values remained high in this sample:

Subscale	Cronbach's α
Inclusion	.75
Ideation	.75
Understanding	.64
Treatment	.74

Comments

- The scale is unique in that it deliberately attempts to assess the interactions among diverse organization members without much reference to race, ethnicity, or culture in the phrasing of the items.

- The development of this instrument was characterized by serious attention to psychometric properties.

- The authors suggest complementing the use of the WDQ with open-ended questions about what specific differences affect group interactions.

Bibliography (studies that have used the measure)

Contact info & cost

Linda K. Larkey
Director, Women's Cancer Prevention Office
300 N. 18th Street
Phoenix, AZ 85006, USA
Tel: 602-462-1005

e-mail: larkeylite@msn.com

General Diversity Measures, Diversity Climate, Multiple Isms

TITLE OF MEASURE	QUICK DISCRIMINATION INDEX
Source/Primary reference	Ponterotto, J. G., Burkard, A., Rieger, B. P. (1995). Development and initial validation of the Quick Discrimination Index (QDI). *Educational and Psychological Measurement, 55,* 1016-1031.
Construct measured	Attitudes toward racial diversity and women's equality
Brief description	The QDI includes 30 items in three domains: 1. Attitudes about diversity 2. Personal attitudes about racial diversity 3. Gender-based attitudes Items are rated on a 5-point scale from 1 = strongly disagree to 5 = strongly agree.
Sample items	Attitudes about diversity: - I am against affirmative action programs in business. Personal attitudes about racial diversity: - Most of my close friends are from my own racial group. Gender-based attitudes: - I think it is more appropriate for the mother of a newborn baby, rather than the father, to stay home with the baby (not work) during the first year.
Appropriate for whom (i.e. which population/s)	Late adolescent and general adult populations
Translations & cultural adaptations available	None known
How developed	Items were generated from the literature on discrimination, prejudice, and "modern racism," and from the development team's experience. An attempt was made to tap both cognitive and affective components of prejudicial attitudes. About 40 statements were initially written. Each item statement was examined by the research team and redundant, unclear, and confusing items were eliminated. Twenty-eight remaining items were placed on a 5-point Likert-type scale. Based on the results of the item content analysis (described below) and factor analysis, two revisions were made. First, the two items having low item-total correlations were closely examined and rewritten. Second, five new items were written, bringing the total item count to 30.

General Diversity Measures, Diversity Climate, Multiple Isms

TITLE OF MEASURE | QUICK DISCRIMINATION INDEX

Psychometric properties

<u>STUDY SAMPLES</u>

Study 1: Undergraduate and graduate students, local church members, and employees of local businesses, human service agencies, and a police precinct

Study 2: Involved two samples:

- Sample 1: Similar to Study 1; late adolescents and adults in the New York metropolitan area

- Sample 2: 37 college undergraduates at a midsize liberal arts college in the Northeastern U.S.

Study 3: Similar to Study 1; late adolescents and adults in the New York metropolitan area

Participants		*Study 1*	*Study 2*		*Study 3*
Sample		Overall	Sample 1	Sample 2	Overall
Sample Size		$n = 285$	$n = 220$	$n = 31$	$n = 331$
Age	Range	18-66	16-58	17-50	16-63
	M (SD)	30 (10)	22 (9)	23 (7)	27 (10)
Gender	Female	187	(59%)	16 (55%)	(79%)
	Male	97	(41%)	15 (45%)	(21%)
Race/Ethnicity	Caucasian	66%	60%	†	76%
	African American	21%	10%		5%
	Hispanic Latino Latina	6%	23%		8%
	Asian American	3%	4%		5%
	Native American	1%	0%		0%
	Other	3%	4%		6%

†Not reported

<u>VALIDITY</u>

Content Validity

To address the possible effects of social desirability, about one-half of the items were written in reverse order. Second, the title "Social Attitude Survey" (not "Quick Discrimination Index") appears on the actual instrument to control somewhat for potential demand characteristics and evaluation apprehension.

General Diversity Measures, Diversity Climate, Multiple Isms

TITLE OF MEASURE   QUICK DISCRIMINATION INDEX

Study 1: Five individuals with expertise in the topical area and in psychological measurement who were not part of the development team rated each item in the prototype QDI on domain appropriateness and clarity. Items receiving a mean of less than 4.0 for either rating (on a 1-5 scale, with 5 indicating highly appropriate or very clear) were eliminated or rewritten. This procedure resulted in a version with 25 items.

The 25-item QDI was then the subject of a 2-hour focus group conducted by the senior author with seven graduate students in education. Focus group members completed the instrument and then discussed their reactions (both affective and cognitive) to the items, satisfying the authors that the items were clear and served their intended purpose. Focus group members completed the QDI in 6 to 13 minutes.

Study 3: A sub-sample of participants ($n = 151$) completed both the QDI and the Social Desirability Scale (SDS; Crowne & Marlowe, 1960). The correlations between the QDI factors and the SDS were low, indicating that social desirability contamination was not a concern (see table in Concurrent Validity).

Construct Validity

Study 1: Using one-way ANOVA, higher scores were found among women than men; non-Whites (all groups combined) than whites; urban dwellers than suburban or rural; and Democrats than Independents than Republicans.

Study 2: In one-way MANOVAs and ANOVAs, scores were higher among African and Hispanic Americans than white Americans, as well as among women, urban dwellers, and Democrats.

Study 2: In factor analysis, Factor 1 accounted for 25.2% of the variance and loaded highly on nine items, all consistent with general/cognitive attitudes toward multicultural issues. Factor 2 loaded highly on seven items, focusing on more personal/affective attitudes toward racial diversity. Factor 3 loaded highly on seven items concerning women's equality attitudes. Factor inter-correlations were moderate and supported the factor extraction. Also presented below are the Cronbach's α values for Study 2 and Study 3.

Factor	Factor 1 $r =$	Factor 2 $r =$	Factor 3 $r =$	Study 2 $\alpha =$	Study 3 $\alpha =$
1		.41**	.47**	.80	.85
2			.35**	.83	.83
3				.76	.65
Total	.83**	.72**	.74**	.89	.88

**$p < .01$

General Diversity Measures, Diversity Climate, Multiple Isms

TITLE OF MEASURE   QUICK DISCRIMINATION INDEX

Confirmatory factor analysis, using structural equation modeling, confirmed the factor structure.

Concurrent Validity

Study 3: A sub-sample of participants ($n = 151$) completed the QDI plus one of two other instruments: the New Racism Scale (NRS; Jacobson, 1985), or the Multicultural Counseling Awareness Scale (MCAS; Ponterotto et al., 1993).

As expected, the NRS was significantly correlated with all three QDI factors but more highly with Factor 1 and Factor 2 (dealing with race) than with Factor 3 (dealing with women's issues). The correlations with the MCAS factors were generally in the same range.

Scale	Factor 1 r =	Factor 2 r =	Factor 3 r =
NRS	.44**	.44**	.30**
MCAS Knowledge & Skills	.41**	.34*	.23
MCAS Awareness	.50**	.21	.39**
SDS	-.16	-.04	-.19

*$p < .05$; **$p < .01$

RELIABILITY

Internal Consistency

Study 1: With the exception of two items, corrected item-total correlations ranged from .20 to .74. The coefficient of variation was 13.4%, within the recommended range of Dawis (1987).

Study 2: Twenty-seven of the 30 items had corrected item-total correlations in the .23 to .62 range. The coefficient of variation was 12.8%. The QDI retained a satisfactory level of internal consistency (see Cronbach's α values above) despite a more diverse developmental sample.

Variable	Study 1	Study 2
# of items	25	30
Mean Corrected Item-total r =	.45	.42

Test-retest Reliability

Study 2: Three college professors distributed the QDI in class during the first and last weeks of the semester. The interval for all three classes was 15 weeks. Test-retest reliability over this interval was high.

General Diversity Measures, Diversity Climate, Multiple Isms

TITLE OF MEASURE   QUICK DISCRIMINATION INDEX

	Mean Stability Coefficient		
Scale	Factor 1	Factor 2	Factor 3
QDI test-retest	.90	.82	.81

Comments

Bibliography (studies that have used the measure)

Contact Information

Joseph G. Ponterotto
Counseling Psychology Program
Graduate School of Education
Fordham University at Lincoln Center
113 W. 60th St, Room 1016A
New York, NY 10023, USA

Tel: 212-636-6480

Race, Racism, Ethnicity, Racial Discrimination & Related Measures

Race, Racism, Ethnicity, Racial Discrimination & Related Measures

TITLE OF MEASURE	**INSTITUTIONAL RACISM SCALE (IRS)**
Source/Primary reference	Barbarin, O. A., & Gilbert, R. (1981). Institutional racism scale: Assessing self and organizational attributes. In O. A. Barbarin, P. R. Good, O. M. Pharr, & J. Siskind (Eds.), *Institutional Racism and Community Competence* (pp.147-171). Rockville, MD: U.S. Department of Health and Human Services.
Construct measured	How individuals construe institutional racism, engage in anti-racism, and view organizational commitment to the reduction of racism
Brief description	The IRS consists of four subscales that assess self attributes and two subscales that assess organizational attributes.

The self-attribute subscales include:

1. Indices of Racism subscale. This includes 8 items frequently cited in the literature as racist to assess individual sensitivity to racism. Respondents rate the items on the extent to which they believe the items are an indication of institutional racism (1 = not at all, 7 = most sensitive).

2. Use of Strategies for Reducing Racism subscale. This subscale includes 7 interventions such as voting, litigation, educating friends, lobbying, and cross-racial interaction. Respondents are first asked to indicate on a four-point scale the effectiveness of the intervention (poor to excellent) and then the extent to which they have personally used these strategies for the purpose of reducing racism (1 = never, 5 = very frequently).

3. Effectiveness of Strategies for Reducing Racism subscale. This subscale consists of 11 items that are similar to the above use items but the items are rated in relation to their effectiveness in reducing racism (1 = poor, 4 = excellent).

4. Personal Efforts to Reduce Racism subscale. This subscale consists of 20 semantic differential ratings regarding how active and how favorably the respondents perceive themselves in reducing racism.

The organizational attribute subscales include:

Agency Climate subscale. This subscale consists of 6 statements related to the extent to which the organizational policies and climate incorporate a respect for minorities and cultural diversity, e.g., in terms of interpersonal processes, decision-making processes, and reward system/ career development processes.

Management/Administrative Efforts to Reduce Racism subscale. This includes 20 semantic-differential ratings to describe how workers perceive management's efforts to reduce racism.

Race, Racism, Ethnicity, Racial Discrimination, & Related Measures

TITLE OF MEASURE: **INSTITUTIONAL RACISM SCALE (IRS)**

Sample items	Indices of Racism: • Personnel selection based on written tests. • Seniority as a major criterion for promotion. Organizational Attributes: • Minority groups have little to say about decisions which affect functioning in this agency. • The organization goes out of its way to make minorities feel at home.
Appropriate for whom (i.e. which population/s)	Adults
Translations & cultural adaptations available	None known
How developed	The initial IRS items were developed based on a literature review of dimensions of institutional racism. Administration of the original IRS raised questions about the suitability and wording of some of the items as well as about the length of time required to complete the questionnaire. The IRS was refined on the basis of Pearson's test-retest and Kuder Richardson's reliability tests. Items with a test-retest reliability of r = .30 or less or intercorrelation with their test mean of .15 or less were eliminated from their respective subscales. The final instrument consists of 72 items – 35 items were eliminated from the original IRS.
Psychometric properties	**STUDY SAMPLE** The IRS was administered to three separate reference groups: Group 1: A conference group. The first group consisted of 56 individuals from educational, religious, and mental health agencies who attended a three-day conference on institutional racism. The participants were given the IRS questionnaire both before and after the conference. A three-day interval separated the administration of pre- and post-tests. The IRS was a part of the battery used to evaluate the impact of the conference. Group 2: A government group. This group included the employees of a single Federal agency (N was not given). The IRS questionnaire was administered on an individual basis with 2-3 month intervals separating pre- and post-test administrations. Also, this government group was asked to indicate examples of racism which had occurred at their workplaces and strategies for the reduction of racism which could potentially serve as measures of institutional racism against which IRS subscales could be compared. Group 3: A student group. The student group included 48 students enrolled in an undergraduate community psychology class who participated in the study for course credit. A 2-month interval separated

Race, Racism, Ethnicity, Racial Discrimination & Related Measures
TITLE OF MEASURE   INSTITUTIONAL RACISM SCALE (IRS)

the administration of pre- and post-tests. The questionnaire was designed to solicit comments about perceived weaknesses and/or ambiguities of the measures, as well as the extent to which the IRS had influenced further thinking about institutional racism.

Participants in all three groups were classified as minorities when they identified themselves as Afro-American, Asian-American, American Indian, or Latinos. Most of the minorities in the sample were Afro-American.

VALIDITY

Concurrent Validity

Pearson correlation coefficients of the IRS subscales among Group 2.

IRS subscales	Spontaneous report of racist incidents	Identification of racist practice from a list	Hope for change
Indices of racism (n = 25)	.33	.51**	-.20
Effectiveness of strategies (n = 21)	.10	.22	-.21
Use of strategies (n = 19)[a]	.38	-.4**	.41*
Personal efforts (n = 24)[a]	.16	.3	-.39*
Agency policies (n = 25)	.53**	.44*	.32
Administrative efforts (n = 19)	-.43*	-.26	.51*

[a] n in each cell varies slightly due to missing values
*$p \leq .05$ **$p \leq .01$

RELIABILITY

Test-Retest and Internal Consistency

The IRS subscales' Test-Retest and KR-14 internal consistency correlations by racial and three reference groups

Race, Racism, Ethnicity, Racial Discrimination, & Related Measures

TITLE OF MEASURE — INSTITUTIONAL RACISM SCALE (IRS)

IRS Subscale	Test-retest correlations				
	Minority	Non-minority	Group 1 (conference)	Group 2 (government)	Group 3 (students)
Indices of racism (n)	.39 (16)	.6 (48)	.5 (8)	.72 (23)	.52 (37)
Effectiveness of strategies (n)	.66 (13)	.81 (42)	.87 (7)	.65 (19)	.79 (32)
Use of strategies (n)	.79 (9)	.69 (27)	.35 (9)	.7 (18)	.55 (11)
Personal efforts (n)	.85 (11)	.74 (42)	.98 (7)	.8 (17)	.7 (22)
Agency policies (n)	.61 (13)	.66 (39)	.95 (6)	.71 (19)	.60 (31)
Administrative efforts (n)	.72 (8)	.73 (37)	.97 (5)	.85 (19)	.69 (25)

IRS Subscale	Internal consistency correlations				
	Minority	Non-minority	Group 1 (conference)	Group 2 (government)	Group 3 (students)
Indices of racism (n)	na	na	na	na	na
Effectiveness of strategies (n)	.56 (45)	.56 (71)	.57 (49)	.26 (41)	.64 (41)
Use of strategies (n)	.69 (40)	.73 (60)	.55 (44)	.48 (34)	.76 (35)
Personal efforts (n)	.83 (42)	.91 (65)	.85 (39)	.88 (41)	.91 (37)
Agency policies (n)	.79 (37)	.72 (64)	.86 (38)	.69 (37)	.73 (38)
Administrative efforts (n)	.93 (33)	.94 (66)	.95 (35)	.91 (39)	.95 (36)

Comments

- The IRS seems to be a reliable measure of individual construction of racism, strategies for altering racist practices, and perceptions of agency climate.

- The IRS subscales seem to have good reliability and their validity is supported by their strong relationship with other measures of institutional racism.

- The IRS concentrates mostly on processes rather than outputs. Also, it provides global ratings that may require the addition of agency-specific items before recommendations for change can be developed.

- The full scale is available in the book chapter.

Race, Racism, Ethnicity, Racial Discrimination & Related Measures

TITLE OF MEASURE	INSTITUTIONAL RACISM SCALE (IRS)
	• The post-questionnaire responses of the conference group participants may have been influenced by the conference itself, therefore affecting test-retest reliabilities. This group had only a 3-day interval between pre- and post-test administration whereas the two other reference groups had a 2-month interval.
Bibliography (studies that have used the measure)	Jeanquart-Barone, S., & Sekaran, U. (1996) Institutional racism: An empirical study. *Journal of Social Psychology, 136*(4), 477-482. Watts, R. J., & Carter, R. T. (1991). Psychological aspects of racism in organizations. *Group and Organization Management, 16*(3), 328-345.
Contact Information	Oscar A. Barbarin, Ph.D. L. Richardson and Emily Preyer Bicentennial Distinguished Professor for Strengthening Families University of North Carolina School of Social Work Chapel Hill, NC 27599-3550, USA Tel: 919 962-6405 Fax: 919 968-4033 Fellow, Frank Porter Graham Child Development Institute Cheryl Mar North Chapel Hill, NC 27599-8040, USA Tel: 919 843-6469 e-mail: barbarin@e-mail.unc.edu

Race, Racism, Ethnicity, Racial Discrimination, & Related Measures

Title of measure	**Hispanic Stress Inventory (HSI)**
Source/Primary reference	Cervantes, R. C., Padilla, A. M., & Salgado de Snyder, N. (1991). The Hispanic Stress Inventory: A culturally relevant approach to psychosocial assessment. *Journal of Consulting and Clinical Psychology, 3*(3), 438-447.
Construct measured	Five psychological stressors in the Hispanic population: marital stress, family stress, occupational/economic stress, discrimination stress, and acculturation stress
Brief description	There are two versions of the scale: one for immigrant Hispanics and one for U.S.-born Hispanics. Immigrant Version 73 items (yes/no, then not at all stressful-extremely stressful) 5 Subscales: 1. Occupational/Economic Stress 2. Parental Stress 3. Marital Stress 4. Immigration Stress 5. Family/Cultural Stress U.S.-Born Version 59 items (yes/no, then not at all stressful-extremely stressful) 4 Subscales: 1. Occupational/Economic Stress 2. Parental Stress 3. Marital Stress 4. Family/Cultural Stress Participants respond by rating whether they have experienced the situation described in each item during the past three months. If affirmative, the degree of stress that the participant associates with the corresponding item is to be rated on a 5-point Likert scale (1 = not at all stressful to 5 = extremely stressful).
Sample items	<ul><li>Since I'm Latino, I felt isolated at work.</li><li>Boss thought I was too passive.</li><li>Others worried about amount/quality of work I do.</li></ul>
Appropriate for whom (i.e. which population/s)	Hispanic adults living in the US
Translations & cultural adaptations available	Spanish translation; there is an 11-item variation developed for Mexican women (Salgado de Snyder, 1987).

Race, Racism, Ethnicity, Racial Discrimination & Related Measures

TITLE OF MEASURE   HISPANIC STRESS INVENTORY (HSI)

How developed	Phase 1: Semi-structured interview consisted of 33 open-ended questions around six psychological stress domains: marital stress, family stress, occupational stress, economic stress, discrimination stress, and acculturation stress. Interviews were tape-recorded. Commonly reported stressor events were identified, developed into short statements, and included in the initial 176 HSI items.
	Phase 2: Five Hispanic judges – two women and three men – rated all 176 items and categorized them into six conceptually meaningful stressor categories (cultural, marital, familial, occupational, economic, and discrimination). Items could be assigned to as many categories as needed to allow overlapping. The judges were asked to pay attention to any awkwardly worded items or items that were irrelevant for the population. Through this process, a new refined HSI, with 133 items in five subscales, was developed.
	Phase 3: The HSI developed in Phase 2 was administered to 493 people. The scale was refined further. If a particular item was reported to be experienced by less than 5% of immigrants, that item was deleted and not included in the further analysis. Items with means less than 2.0 (somewhat stressful) were also deleted. The construct validity and correlations with other pre-selected measures were examined. Two HSI versions were established: (i) for immigrant Hispanics, and (ii) for U.S.-born Hispanics.
	Phase 4: Reliability estimates – estimates of internal consistency and test-retest procedure – were obtained.
Psychometric properties	**STUDY SAMPLES**:

Participants		Phase 1	
		Group 1	*Group 2*
Sample Size		n = 43	n = 62
Description		convenience sample of 43 Hispanic adults	sample of 62 Mexican and Central American adults who were more recent immigrants - these respondents had 5 years of residency in the US
Age	Range Mean	20-69	
	Mean	39	33.8
Gender	Female	44.2%	50.0%
	Male	55.8%	50.0%
Ethnicity	Mexican born	53.5%	
	Mexican-American	46.5%	
	Mexican		51.6%
	Salvadoran		27.4%
	Guatemalan		17.7%
	Honduran		1.6%
	Nicaraguan		1.6%

Race, Racism, Ethnicity, Racial Discrimination, & Related Measures

TITLE OF MEASURE   HISPANIC STRESS INVENTORY (HSI)

Participants		Phase 3	
Sample Size		*n*= 493	*n*=141
Description		Volunteers from two adult community schools located in the Central Downtown area of Los Angeles and East Los Angeles	To ensure cultural specificity of the HAS fir Hispanic culture, items common to the Immigrant and U.S.-Born versions were administered to a <u>non-Hispanic sample</u>.
Age	Range	17-56	17-40
	Mean	23	22
Gender	Female	*n*=238 (48.3%)	*n*=78 (55.3%)
	Male	*n*=255 (51.7%)	*n*=63 (44.7%)
Education		13 years	13 years
Per Capita Income		$464 per month	$1,147 per month
Employed		40.9%	66%
Marital Status	Married	13.6%	11%
	Single	79.6%	-
	Divorced/Separated	4.0%	-
Number of Children (mean)		1.9	-
Number of Persons Living at Home (Mean)		4.9	-
Ethnicity	Mexican-born	3.2%	-
	Central American	24.3%	-
	Other Latin American	9.5%	-
	U.S.-born	38.1%	-
	Anglo-American		100%

VALIDITY

Content Validity

Phase 1: Commonly reported stressor events were identified through interviews. The initial 176 HSI items were selected and developed into a series of short statements.

Phase 2: Five Hispanic judges – two women and three men – rated all 176 items and categorized them into six conceptually meaningful stressor categories (cultural, marital, familial, occupational, economic, and discrimination). The judges reached complete agreement on the categorization of 79 items (45%). Four of five judges were in agreement on the assignment of an additional 52 items (30%). If an item was not categorized similarly by at least four of five judges, it was discarded, unless the item was seen as clinically important by the authors. Individual items were also removed if three of five judges thought the item to be unrelated to stress. The remaining 133 items were randomly ordered, producing a refined first version of HSI with five subscales; occupational and economic scales were combined.

Race, Racism, Ethnicity, Racial Discrimination & Related Measures

TITLE OF MEASURE: HISPANIC STRESS INVENTORY (HSI)

Construct Validity

The factor analyses yielded a five-factor solution for the immigrant subsample and a four-factor solution for the U.S.-born subsample. The average factor loading for the immigrant sample was .55 and for the U.S.-born sample it was .56.

Final factor solutions and influence on the total variance

HSI Factors	% of total variance
Immigrant Version subscales	
Occupational/Economic Stress	13
Parental Stress	8
Marital Stress	6
Immigration Stress	5
Family/Culture Stress	3
U.S.-Born Version subscales	
Marital Stress	13
Occupational/Economic Stress	10
Parental Stress	7
Family Culture Stress	6

To ensure the cultural specificity of the HSI to Hispanic culture, items common to both the immigrant and U.S.-born versions of the HSI were administered to the combined Hispanic sample (immigrant and U.S.-born), as well as to a non-Hispanic, Anglo-American sample. Factor analysis of the item responses of the combined Hispanic sample yielded five conceptually distinct factors, with the first factor accounting for 12% of the total variance, followed by 8%, 6%, 4%, and 3% for the remaining four factors. In contrast, comparative factor analysis of the item responses of the non-Hispanic sample yielded 12 factors with eigenvalues greater than one. Next, a five-factor extraction was performed on the item responses of the non-Hispanic sample for more direct comparison with the structure that emerged for the combined Hispanic sample. Results of this extraction demonstrated that the first factor accounted for a large percentage (29%) of the total variance and the remaining four factors accounted for an additional 17% of the variance. Further, the factors that did emerge were not interpretable. Thus, the factor matrix of item responses of the non-Hispanic sample differed markedly from that of the combined Hispanic sample, indicating the specificity of the HSI to Hispanic culture.

Pearson correlations were computed to examine the strength of the relationships between each of the HSI subscale scores and the pre-selected criterion measures: SCL-90-R = Symptom Checklist-90-

Race, Racism, Ethnicity, Racial Discrimination, & Related Measures

TITLE OF MEASURE   HISPANIC STRESS INVENTORY (HSI)

Revised (Derogatis, 1977); CES-D = Center for Epidemiological Studies Depression Scale (Radloff, 1977); RSI = Rosenberg Self-Esteem Inventory (Rosenberg, 1965); PCI = Campbell Personal Competence Inventory (Campbell, Converse, Miller, & Stokes, 1960).

HSI Subscales	Symptomatology SCL-90-R		
	Somatization	Depression	Anxiety
Immigrant Version (*n* = 305)			
Occupational/Economic Stress	.21***	.26	.17
Parental Stress	.04	.11	.06
Marital Stress	.16	.20***	.17
Immigration Stress	.20***	.26***	.17
Family/Culture Stress	.30***	.36***	.31***
U.S.-Born Version (*n* = 188)			
Marital Stress	.12	.19	.19
Occupational/Economic Stress	.11	.22	.17
Parental Stress	.05	.07	.04
Family Culture Stress	.29***	.38***	.34***

***$p < .001$

HSI Subscales	Symptomatology		
	CES-D	RSI	PCI
Immigrant Version (*n* = 305)			
Occupational/Economic Stress	.23***	-.15	.11
Parental Stress	.12	-.07	-.04
Marital Stress	.25***	-.06	-.02
Immigration Stress	.27***	-.10	.06
Family/Culture Stress	.45***	-.18***	-.04
U.S.-Born Version (*n* = 188)			
Marital Stress	.17	-.06	.14
Occupational/Economic Stress	.31***	-.07	.03
Parental Stress	.10	-.01	-.03
Family Culture Stress	.40***	-.17	-.03

***$p < .001$

RELIABILITY

Internal Consistency and Reliability

Coefficient alphas were obtained for the Phase 3 data set. For the test-retest study, participants were 35 adult members of a local church group born either in Mexico or Central America.

Race, Racism, Ethnicity, Racial Discrimination & Related Measures

TITLE OF MEASURE   HISPANIC STRESS INVENTORY (HSI)

HSI subscale internal consistencies and reliabilities.

HSI subscales	Coefficient α	Test-retest Pearson coefficient
Immigrant Version subscales		
Occupational/Economic Stress	.91	.79****
Parental Stress	.88	.73****
Marital Stress	.86	.61****
Immigration Stress	.85	.80****
Family/Culture Stress	.77	.86
U.S.-Born Version subscales		
Marital Stress	.90	-
Occupational/Economic Stress	.88	-
Parental Stress	.85	-
Family Culture Stress	.85	-

****$p < .0001$

Comments

- The measure addresses stresses in multiple domains and could be adapted to be more specific to the work setting.

- While somewhat similar to the FASE (Padilla, Wagatsuma, & Lindholm, 1985), the HSI is specifically designed to tap stressors faced by Hispanic adults.

- Given the systematic approach, it was possible to generate a list of stressors both for recent Hispanic immigrants and for U.S.-born Hispanics. Therefore, two separate versions of the HSI were established. one for immigrants, and one for U.S.-born Hispanics.

- Studies show that the HSI has good reliability and validity.

Bibliography (studies that have used the measure)

Cervantes, R., Padilla, A., & Salgado de Snyder, N. (1990). Reliability and validity of Hispanic Stress Inventory. *Hispanic Journal of Behavioral Sciences, 11*(1), 76-82.

Contact Information

Richard C. Cervantes
Behavioral Assessment, Inc.
291 South La Cienega Blvd., Suite 308
Beverly Hills, CA 90211, USA

Tel: 310-652-6449

Fax: 310-652-5462

e-mail: bassessment@aol.com

web address: www.bai-eval.com/download/rccvitabai2002.pdf

Race, Racism, Ethnicity, Racial Discrimination, & Related Measures

TITLE OF MEASURE	**PERCEPTIONS OF RACISM SCALE (PRS)**
Source/Primary reference	Green, N. L. (1995). Development of the Perceptions of Racism Scale. *Journal of Nursing Scholarship, 2*(2), 141-146.
Construct measured	Perceptions of racism against African Americans
Brief description	The PRS is a 20-item self-report measure of perceived racism. The instrument is a *single*-dimension measure of racism. Each item is rated on a 4-point scale; respondents are asked to indicate the extent to which they agree with a given statement. The scale range is from 1 = strongly agree to 4 = strongly disagree. A high score indicates high perceptions of racism.
Sample items	African American women experience negative attitudes when they go to a white doctor's office. - Racism is a problem in my life. - Officials listen more to whites than African Americans.
Appropriate for whom (i.e. which population/s)	Adults
Translations & cultural adaptations available	None known
How developed	The PRS items were developed on the assumption that racism perceptions can be divided into three categories: affective, behavioral, and cognitive. Items were collected from two sources: 1) interviews with 8 African American childbearing women about their perceptions of racism; and 2) a Business Week/Harris Poll regarding perceptions of general racism (employment, housing, judicial system). The items were ordered to mix health and general statements and to allow reversal statements and a mix of positive and negative responses. The items were reviewed by six African American nurse-midwives and one teacher (content validity). Two experts in instrument design judged item syntax. The selected items were duplicated and reversed. The result was the final 20-item instrument. The initial conceptualization of PRS distinguished two subscales representing 1) health care and 2) societal racism.

Race, Racism, Ethnicity, Racial Discrimination & Related Measures

TITLE OF MEASURE   PERCEPTIONS OF RACISM SCALE (PRS)

Psychometric properties

STUDY SAMPLES

Participants		Study 1	Study 2
Sample Size		n = 109	n = 136
Description (Convenience Samples)		African American educated, employed women; churches & community organizations	African American pregnant women; health clinic
Age	Range	20-80	18-39
	Mean (SD)	47 (15)	24 (5)
Education	Range	2-18 years	8-18 years
	Mean (SD)	15 (3)	14 (2)
Monthly Family Income (categorized)	Range	$500 or less – over $2,600	$500 or less – over $4,000
	Mean	$1,701-$2,000	$1,501-$2,000
	Mode	over $2,600	$1,501-$2,000
	Median	$2,201-$2,600	$1,501-$2,000
Marital Status	Never Married	12 (11%)	81 (60%)
	Married	51 (47%)	47 (35%)
	Widowed	14 (13%)	0
	Separated/Divorced	30 (27%)	5 (4%)
	Other	0	3 (2%)
	Missing	2 (2%)	0

VALIDITY

Content Validity

Content validity was assessed by asking 6 African American nurse-midwives and one teacher to provide written and verbal critiques of the assumption that the scale content had been adequately sampled and translated into scale items.

Construct Validity

In study 1, an orthogonal rotation did not result in any clear division of the items into two separated subscales. Principal components analysis with rotation resulted in a single factor accounting for 41% of the total variance. As a result, the 20 items were retained in a single scale.

In study 2, a lower perception of racism was found. Responses were significantly different between the two groups on all items except two. Responses to the two items were not correlated with the overall responses.

Race, Racism, Ethnicity, Racial Discrimination, & Related Measures

TITLE OF MEASURE	PERCEPTIONS OF RACISM SCALE (PRS)
	<u>RELIABILITY</u> ***Internal Consistency*** Cronbach's α coefficients were high in each pilot study. | Scale	Study 1 α =	Study 2 α =
---	---	---
PRS	.86	.91	
Comments	<ul><li>The author mentioned that a hypothesized positive relationship between racism and stress was found, but specific evidence of relationship to health was not presented.</li><li>The scale is a unidimensional measure of racism. However, given the ample evidence of racism's multidimensional nature, it is unclear what dimension PRS actually captures (i.e., unlikely a measure of the full experience of racism).</li><li>Items were developed based on interviews with childbearing African American women and some items are very specific to interactions with medical providers. While many general racism items are included, the scale may be particularly useful for assessing perceptions of racism in health care.</li><li>Although many of the items would seem transferable to other groups in other situations, the scale's usefulness with a broader African American population or non-African American populations is unknown.</li></ul>
Bibliography (studies that have used the measure)	Murrell, N. (1996). Stress, self-esteem, and racism: Relationships with low birth weight and preterm delivery in African American women. *Journal of National Black Nurses Association, 8*(1), 45-53.
Contact Information	Nanny Green University of California San Francisco, CA, USA

Race, Racism, Ethnicity, Racial Discrimination & Related Measures

TITLE OF MEASURE	THE RACISM AND LIFE EXPERIENCE SCALES (RaLES)
Source/Primary reference	Harrell, S. P. (1997). *The Racism and Life Experience Scales.* Unpublished instrument. Los Angeles, CA: Pepperdine University Graduate School of Education and Psychology.
	Harrell, S. P., Merchant, M. A., & Young, S. A. (August, 1997). *Psychometric properties of the Racism and Life Experiences Scales (RaLES).* Presented at the annual meeting of the American Psychological Association. Chicago, IL.
Construct measured	The RaLES is a comprehensive set of scales that measures racism-related stress, coping, socialization, and attitudes. Only the scales for frequency and stressfulness of racism-related experiences are described here.
Brief description	The RaLES includes five primary scales and one overview scale that assess the frequency, intensity, and stressfulness of multiple dimensions of racism-related experiences. (The ratings for each set of questions are listed in the next section.)

1. The *Perceived Influence of Race (PER)* scale assesses the degree to which race is judged to have influenced one's life experiences across twenty contexts of living (finding a job, quality of education, family life, money and finances, health, etc.). It reflects stress that is chronic, contextual, or role-related (vs. specific life events).

2. The *Racism Experiences (EXP)* scales assess the frequency of 17 specific types *(EXP-TP)* of direct and vicarious experiences of racism over a specified time period (e.g., past month, year, 3 years, lifetime), as well as the stressfulness of those experiences *(EXP-ST)* and the domains of daily life *(EXP-DM)* in which they have occurred (ten areas of life such as employment, financial, and health care).

3. The *Daily Life Experiences (DLE)* scale assesses the frequency, perceived involvement of race, and stressfulness of 20 daily "micro-experiences" (i.e., daily hassles) over a specified period of time. Three subscales (frequency, race involvement, and bother) are summed for the total score.

4. The *Life Experiences and Stress (STR)* scale is a comprehensive inventory of the occurrence and stressfulness of 128 specific personal life events within 9 life contexts (e.g., employment, community life, law enforcement and legal system). This scale can be administered in full (128 items), or specific contexts can be chosen as relevant. The items include both generic stressors and stressors associated with discrimination.

5. The *Group Impact (GRP)* scale assesses collective racism experiences, those that involve the observation of racism in the lives of others of one's own group regardless of personal experience.

Race, Racism, Ethnicity, Racial Discrimination, & Related Measures

TITLE OF MEASURE	THE RACISM AND LIFE EXPERIENCE SCALES (RaLES)

The scale includes 16 areas of life such as employment, education, housing, and health care/health status.

The *Brief* scale is a general overview measure of racism-related stress that may be used as an alternative to the full instrument. It includes 9 questions that assess direct, vicarious, and collective experiences of racism, as well as the stressfulness of racism.

Sample items

Perceived Influence of Race (PER)

- How much do you think that your race has influenced your life in the area of experiences at your job?

5-point Likert scale ranging from 0 = Not at all influenced by my race to 4 = extremely influenced by my race.

Domains of Racism Experience (EXP-DOM)

How much have you personally experienced racism, racial discrimination, or racial prejudice during the past 1 year (may vary) in each of the following areas of your life:

- Employment/job
- Loans, credit, financial matters

5-point Likert-type scale ranging from 0 = not at all to 4 = extremely.

Racism Experiences (EXP-TP, EXP-ST)

Listed below are different types of racism-related experiences that some people have. Please think about experiences you might have had involving racism, racial discrimination, or racial prejudice during the past year (may vary) and rate how often you had the experience and how stressful the experience was:

- Conflict between you and someone of a different race/ethnicity
- Witnessing discrimination or prejudice directed toward someone else
- Hearing about someone else's experience of discrimination or prejudice
- Observing limited participation in decision-making, opportunities, access to resources for people of your racial/ethnic group (i.e., "ol' boys network")

Race, Racism, Ethnicity, Racial Discrimination & Related Measures

TITLE OF MEASURE — THE RACISM AND LIFE EXPERIENCE SCALES (RaLES)

5-point Likert-type scales: frequency responses range from 0 = never to 4 = very often; stress-bother responses range from 0 = has never happened to me to 4 = extremely.

Daily Life Experiences (RaLES-DLE)

These questions ask you to think about experiences that some people have as they go about their daily lives. Think only about the past year (may vary). Please consider how often you usually have each of the experiences listed below:

- Others reacting to you as if they were afraid or intimidated
- Hearing or being told an offensive joke
- Others expecting your work to be inferior
- Being mistaken for someone who serves others (e.g., janitor, maid, etc.)
- Being asked to speak for or represent your entire racial/ethnic group (e.g., "What do _____ people think?")

6-point Likert scales: frequency responses range from 0 = never to 5 = once a week or more; stress-bother responses range from 0 = has never happened to me to 5 = bothers me extremely.

Life Experiences and Stress (RaLES-STR)

EMPLOYMENT. Think about your experiences related to employment and the jobs you have had. Place a check by any experience that has ever been a problem for you. Then, only for the ones that you checked, answer whether or not racism has been involved in the difficulties that you have had:

- Deciding on a career to pursue
- Not having a mentor or someone to "show you the ropes"
- Being assigned undesirable tasks or projects at a job
- Taking a job below your abilities or education
- Not receiving information or communication (being left "out of the loop")
- Having your work criticized frequently or being watched closely at your job

Two additional items are included for each life context (e.g., employment) concerning stressfulness of that context in the past year and during one's lifetime.

Race, Racism, Ethnicity, Racial Discrimination, & Related Measures

TITLE OF MEASURE	THE RACISM AND LIFE EXPERIENCE SCALES (RaLES)

Group Impact (GRP)

Please indicate how much you believe racism affects each of the following areas of life for people of your same racial/ethnic group, even if your personal experiences have not been related to racism

- Things that happen in the workplace or related to employment
- Things that happen in schools and the educational system
- Health status and health care
- Relationships between people of your same racial/ethnic group

5-point Likert-type scale ranging from 0 = not at all influenced by racism to 4 = extremely influenced by racism.

Brief scale (RaLES-B)

- DURING THE PAST YEAR, how much have you personally experienced racism, racial discrimination, or racial prejudice?
- Overall, how much do you think racism affects the lives of people of your same racial/ethnic group?
- In general, how frequently do you hear about incidents of racial prejudice, discrimination, or racism from family, friends, co-workers, neighbors, etc.?

All items rated on 5-point Likert-type scales.

Appropriate for whom (i.e. which population/s)	Adolescents and adults of diverse racial/ethnic heritage. Most appropriate for members of historically oppressed racial/ethnic groups (e.g., African Americans, Native Americans, Latino, Arab/Middle Eastern, etc.)
Translations & cultural adaptations available	None known
How developed	Development of the RaLES began in a substance abuse referral and treatment project among African American and Latino men in Los Angeles in 1991. The author expanded the initial items assessing the frequency and stressfulness of racism-related life events in 1993 and developed the first three scales (PER, GRP, and STR). Scale items were developed based on literature review, focus groups, and the author's experiences. The remaining scales were developed in 1994-1997 and operationalized the conceptualization of the multidimensionality of racism-related stress (see Harrell, 2000).

Race, Racism, Ethnicity, Racial Discrimination & Related Measures

TITLE OF MEASURE: THE RACISM AND LIFE EXPERIENCE SCALES (RaLES)

Psychometric properties

<u>STUDY SAMPLES</u>: Six psychometric studies conducted between 1993-1996 have provided data on the reliability and validity of the RaLES scales.

Sample	Description
Development Sample	Racially and ethnically diverse undergraduate and graduate students from colleges and universities in Los Angeles County
Sample 2	Ethnically diverse college freshmen
Sample 3	Racially and ethnically diverse students in pre-freshman and pre-transfer summer programs at a large West Coast university
Sample 4	Undergraduate and graduate students from the same West Coast university
Sample 5	National sample of African American adults recruited from professional organizations
Sample 6	African American adults recruited from community settings and networks known to the author

Participants		Development Sample	Sample 2	Sample 3
Sample Size		n = 286	n = 126	n = 187
Age	Range	18 – 39+	16 – 39	16 – 40
	Mean	†	†	18.44
Gender	Female	76.5%	65.9%	67.4%
	Male	23.5%	34.1%	32.1%
Race/Ethnicity	African American	15.1%	24.6%	19.4%
	Latino	10.3%	57.9%	62.9%
	Asian/Pacific Islander	9.3%	11.1%	8.6%
	Middle Eastern	4.0%	0%	0.5%
	American Indian	1.5%	0%	†
	Biracial/Multiracial	4.0%	6.3%	5.9%
	White (non-Jewish)	51.5%	0%	1.6%
	White-Jewish	4.0%	0%	†
	Other	†	†	1.1%

†Not reported

Race, Racism, Ethnicity, Racial Discrimination, & Related Measures

Participants		Sample 4	Sample 5	Sample 6
Sample Size		n = 150	n = 104	n = 50
Age	Range	16 – 60+	16 – 60+	†
	Mean	22.18	40.47	†
Gender	Female	62.4%	73.1%	†
	Male	37.6%	26.9%	†
Race/ Ethnicity	African American	26.8%	100%	100%
	Latino	29.5%	-	-
	Asian/Pacific Islander	28.9%	-	-
	Middle Eastern	0.7%	-	-
	American Indian	†	-	-
	Biracial/Multiracial	4.0%	-	-
	White (non-Jewish)	2%	-	-
	White-Jewish	†	-	-
	Other	5.4%	-	-

†Not reported

VALIDITY

Content Validity

In multiple samples, correlations with social desirability were either small or not statistically significant for the PER, EXP-DM, EXP-TP, DLE, and RaLES-B scales.

EXP-DM had a small negative correlation with social desirability in sample 2. In sample 3, DLE-frequency, EXP-TP (direct), and EXP-ST subscales had small, but statistically significant, negative correlations. The GRP scale also had a significant negative correlation with social desirability.

No data were available for the STR scale.

Concurrent Validity

Most of the scales were significantly correlated as expected with criterion measures, indicating strong concurrent validity:

<u>PER</u> with perceived discrimination, racism reaction, urban life stress, and collective self esteem (Samples 1 and 3).

<u>EXP-DM</u> with urban stress, racial discrimination, collective self-esteem, and cultural mistrust.

<u>DLE</u> subscales with collective self-esteem, cultural mistrust, racial discrimination, racism reaction, and urban life stress.

<u>EXP</u> with measures of urban life stress, collective self-esteem, racial discrimination, and cultural mistrust.

Race, Racism, Ethnicity, Racial Discrimination & Related Measures

TITLE OF MEASURE — THE RACISM AND LIFE EXPERIENCE SCALES (RaLES)

GRP with urban life stress, collective self-esteem, racism reaction, and racial discrimination.

RaLES-B with urban life stress, collective self-esteem, racial discrimination, and cultural mistrust.

RELIABILITY

Internal Consistency and Reliability

Across the various population samples, reliability was high or moderately high for the PER; EXP-DM; DLE-frequency, DLE-race involvement, and DLE-stress/bother subscales; EXP frequency and stressfulness subscales (as well as for the direct experiences and vicarious experiences factors that emerged); GRP; and RaLES-B scales. No data were available for the STR scale.

Scale	Sample	Cronbach's α	Split-half reliability	Test-retest reliability
PER	1	.91	.90	.79
	3	.91		
EXP-DM	2	.82		
	3	.84		
	4	.90		
	5	.85		
DLE-frequency	1	.89	.85	.79
	3	.89		
DLE-race	1	.94		
	2	.84		
	3	.92		
	4	.94		
	5	.90		
DLE-bother	4	.94		
	5	.93		
EXP-frequency	2	.83	.82	
	3	.86		
	4	.90		
	5	.88		
EXP-frequency (direct)	3	.74		
	4	.85		
	5	.84		
EXP-frequency (vicarious)	3	.85		
	4	.87		
	5	.83		
EXP-stressfulness	2	.88	.83	
	3	.89		
EXP-stressfulness (direct)	3	.74		
EXP-stressfulness (vicarious)	3	.87		
GRP	1	.96	.94	.86
	3	.92		
RaLES-B	2	.86	.82	
	3	.86		
	4	.77		
	5	.79		

Race, Racism, Ethnicity, Racial Discrimination, & Related Measures

TITLE OF MEASURE | THE RACISM AND LIFE EXPERIENCE SCALES (RaLES)

Comments

- Several of the psychometric studies described above also provided preliminary data on the relationship between the RaLES scales and health outcomes. In Sample 3, the DOM, EXP (direct experiences), and DLE (frequency) had significant negative correlations with positive well-being, while the GRP scale had a significant positive correlation with positive well-being. The DOM, EXP, DLE, and BRF scales were all significantly and positively correlated with psychological symptomatology (i.e., depression, anxiety, somaticization) in Sample 4. In Sample 5, the DOM, EXP, DLE, and BRF scales were all significantly correlated with trauma-related symptoms. In addition, after controlling for demographic variables and experiences of other forms of discrimination (e.g., sexism, classism), the DLE and RaLES-B scales accounted for a significant proportion of explained variance in trauma-related symptoms.

- The strengths of the RaLES include: 1) its comprehensive approach to the measurement of racism experiences and stress; 2) the ability for users to choose one or more scales based on need; 3) ease of administration; 4) applicability to different racial/ethnic groups; and 5) data suggesting strong psychometric properties. A full Interview Version is under development for populations where literacy may be a concern.

- The primary drawbacks of the RaLES include its length, its limited use in published studies to date, and the need to develop norms for broader and more representative samples.

Bibliography (studies that have used the measure)

Harrell, S. P. (2000). A multidimensional conceptualization of racism-related stress: Implications for the well-being of people of color. *American Journal of Orthopsychiatry, 70*, 42-57.

Sellers, R. M., & Shelton, N. J. (2003). The role of racial identity in perceived racial discrimination. *Journal of Personality and Social Psychology, 84*, 1079-1092.

Utsey, S. (1998). Assessing the stressfulness of racism: A review of instrumentation. *Journal of Black Psychology, 24*, 269-288.

The RaLES has been used in numerous doctoral dissertations from the California School of Professional Psychology. These include:

Cotton, L. M. (1999). *The impact of stress, exposure to violence, and racism on HIV knowledge, attitudes, and behaviors.*

Garcia, R. A. (1998). *The role of socialization influences, racism-related stress, and perceptions of collective racism in adopted patterns of acculturation among young adult Mexican Americans.*

Race, Racism, Ethnicity, Racial Discrimination & Related Measures

TITLE OF MEASURE	THE RACISM AND LIFE EXPERIENCE SCALES (RaLES)

Hagen, K. L. (1997). *The impact of child maltreatment experiences, adult revictimization, history of traumatization symptoms, and racism on the psychological well-being of African American women.*

Miller, J. L. (2001). *Understanding achievement attribution and achievement motivation among African American youth: Racism, racial socialization, and spirituality.*

Oh, M. Y. (2001). *Contingencies of self-esteem: Psychological well-being and impact of perceived experiences of discrimination among Korean Americans.*

Rivera, B. C. (1997). *Perceptions of racism, acculturation, and depression in first-generation Mexican American immigrants and descendants of Mexican American immigrants.*

Rosas, M. C. (1999). *The impact of affirmation action legislation and racism experiences on the collective self-esteem and psychological well-being of college students of color.*

Contact Information

Shelly P. Harrell, Ph.D.
Professor of Psychology
Pepperdine University
Graduate School of Education and Psychology
6100 Centre Drive / Howard Hughes Center
Los Angeles, CA 90045, USA

e-mail: sharrell@pepperdine.edu

Race, Racism, Ethnicity, Racial Discrimination, & Related Measures

TITLE OF MEASURE	WORKPLACE RACIAL BIAS
Source/Primary reference	Hughes, D., & Dodge, M. (1997). African American women in the workplace: Relationships between job conditions, racial bias at work, and perceived job quality. *American Journal of Community Psychology, 25*(5), 581-600.
Construct measured	Experiences of interpersonal and institutional discrimination at work
Brief description	The instrument includes 13 items along two dimensions: 1. Institutional discrimination - 5 statements about the extent to which systems-level transactions are biased (e.g. salaries, job assignments, promotions) 2. Interpersonal prejudice - 8 statements about experiences of racial bias in daily interactions (e.g. jokes and slurs, assumption of incompetence) All statements are rated on a 4-point scale from strongly agree to strongly disagree.
Sample items	Institutional Discrimination: • There is discrimination against *[ethnic group]* in salaries. • *[Ethnic group]*s get the least desirable assignments. Interpersonal Prejudice: • People notice your ethnic background before they notice anything else about you. • People you work with have stereotypes about [ethnic group] that affect how they judge you.
Appropriate for whom (i.e. which population/s)	For Institutional Discrimination items, it is appropriate for all working adults. For the Interpersonal Prejudice items, it is most appropriate for workers of color (non-majority workers).
Translations & cultural adaptations available	Spanish translation available
How developed	Items for the two scales were developed based on a series of six focus groups with African American workers in blue and white collar jobs.

Race, Racism, Ethnicity, Racial Discrimination & Related Measures

TITLE OF MEASURE   WORKPLACE RACIAL BIAS

Psychometric properties

STUDY SAMPLE

Participants		Demographics
Sample Size		n = 79
Description		Full-time employed African American women in married-couple families with at least one child between the ages of 4 and 14 years, from 30 different communities.
Age	Range	21-53
	Mean	37
Education	College	22%
	High School	95%
Income	Median Personal	$10,000-$24,999
	Median Family	$40,000-$54,000
Positional Tenure	Mean	7.5 years

The authors have also used the scale in studies with diverse Latino samples (e.g., Enchautegui de Jesus & Hughes, in preparation).

VALIDITY

Construct Validity
Principal axis factor analysis of ratings on all developed items confirmed two distinct dimensions of workplace bias (items loading above .6 on one factor and below .45 on the other were retained).

Concurrent Validity
The measure of institutional discrimination was significantly correlated with a single item assessing discrimination in workers' present jobs ($r = .40$), but not with a similar item assessing discrimination in past jobs. This seems to indicate that the measure assesses current discrimination and not just a predisposition to perceive/report discrimination. The interpersonal discrimination scale was not associated with global items assessing either present or past discrimination.

RELIABILITY

Internal Consistency
Cronbach's α reliability coefficients by group:

Group	Institutional discrimination α =	Interpersonal prejudice α =
Puerto Rican	.90	.84
Dominican	.90	.79
Black	.85	.83
Mexican	.93	.83
Men	.89	.84
Women	.95	.84

Race, Racism, Ethnicity, Racial Discrimination, & Related Measures

TITLE OF MEASURE: **WORKPLACE RACIAL BIAS**

Comments

Specifically designed to assess the work environment.

- Spanish translation is available.
- Relies on respondents' perceptions.
- This instrument is short and easy to administer.
- Appears reliable with multiple ethnic/racial groups.

Bibliography (studies that have used the measure)

Enchautegui de Jesus, N. (2002). Relationships between normative and race/ethnic-related job stressors and marital and individual well-being among Black and Latino/a workers. *Dissertation Abstracts International, 62*(8-B), 3834.

Enchautegui de Jesus, D., & Hughes, D. (in preparation). *Relationships between job discrimination, psychological well-being, and psychological distress among Latino and Black adults.* New York University and University of Michigan.

Hughes, D., & Chen, L. (1997). When and what parents tell children about race: An examination of race-related socialization in African American families. *Applied Developmental Science, 1*(4), 200-214.

Hughes, D., & Chesir-Tehran, D. (in preparation). *Relationships between job characteristics, job discrimination, and the quality of parenting among dual-earner African American families.* New York University.

Contact Information

Dianne Hughes
Department of Psychology
New York University
6 Washington Place
New York, NY 10003, USA

Race, Racism, Ethnicity, Racial Discrimination & Related Measures

TITLE OF MEASURE	**KRIEGER MEASURE OF EXPERIENCES OF DISCRIMINATION**
Source/Primary reference	Krieger, N. (1990). Racial and gender discrimination: Risk factors for high blood pressure? *Social Science Medicine, 30*(12), 1273-1281. Krieger, N., & Sidney S. (1996). Racial discrimination and blood pressure: The CARDIA study of young black and while adults. *American Journal of Public Health, 86,* 1370-1378.
Construct measured	Self-reported experiences of and responses to racial discrimination
Brief description	The instrument first asks respondents about their typical response to unfair treatment and then asks respondents about whether they have ever experienced racial discrimination in seven different domains. It is a self-administered paper-and-pencil instrument.

Sample items	We are going to ask you a number of questions related to discrimination. Please select one response on questions 1 and 2.
	
	1. If you feel you have been treated unfairly, do you usually:
	
	    __ Accept it as a fact of life?
	
	    __ Try to do something about it?
	
	2. And if you have been treated unfairly, do you usually:
	
	    __ Talk to other people about it?
	
	    __ Keep it to yourself?
	
	3. Have you ever experienced discrimination, been prevented from doing something, or been hassled or made to feel inferior in any of the following seven situations because of your <u>race or color</u>?

At school	No___ Yes___
Getting a job	No___ Yes___
At work	No___ Yes___
Getting housing	No___ Yes___
Getting medical care	No___ Yes___
From the police or in the courts	No___ Yes___
On the street or in a public setting	No___ Yes___

Race, Racism, Ethnicity, Racial Discrimination, & Related Measures

TITLE OF MEASURE	**KRIEGER MEASURE OF EXPERIENCES OF DISCRIMINATION**
Appropriate for whom (i.e. which population/s)	Adolescents or adults
Translations & cultural adaptations available	Currently being translated into Spanish and tested among Latinos/Latinas, as part of the validation study now under way (see "Psychometric Properties," below).
How developed	The instrument was developed for the CARDIA (Coronary Artery Risk Development in Young Adults) study. The questions were developed by the author based on a review of the extant literature (on racial discrimination, measurement of social stressors, etc.), plus pilot testing both for the initial study, published in 1990, and then among CARDIA participants, for the 1996 article.
Psychometric properties	The discrimination questions (pertaining to discrimination based on race/ethnicity, gender, social class, sexual orientation, and religion) were pilot tested by CARDIA staff for their acceptability to CARDIA participants. No explicit psychometric evaluation was conducted. Two new developments are: 1. A recently conducted and as-of-yet unpublished analysis, performed as part of a new CARDIA-based study looking at risk of low birth weight in relation to racial discrimination, gave a Cronbach's α for the racial discrimination measure of 0.78. 2. Data collection is under way (2003) for a study to evaluate the validity and reliability of a revised version of the racial discrimination instrument, in a population of working class African Americans and Latinos/Latinas.
Comments	• Used in studies of African Americans (could be adapted for other populations of color) and white Americans, including persons of low literacy and also very low income. • The studies cited below provide evidence on associations with: blood pressure, preterm delivery, self-reported health status, cigarette smoking, and alcohol-related behaviors. • The instrument is concise, easy to understand, and easy to administer. • The instrument does not capture the duration, intensity, or frequency of the self-reported experiences of racial discrimination; it also asks only about the respondent's experiences.
Bibliography (studies that have used the measure)	Broman, C. L. (1996). The health consequences of discrimination: A study of African Americans. *Ethnicity Disease*, 6, 148-152.

Race, Racism, Ethnicity, Racial Discrimination & Related Measures

TITLE OF MEASURE   KRIEGER MEASURE OF EXPERIENCES OF DISCRIMINATION

Broman, C. L., Mavaddat, R., & Hsu, S. (2000). The experiences and consequences of perceived racial discrimination: A study of African Americans. *Journal of Black Psychology, 26,* 165-180.

Collins, J. W., David, R. J., Symons, R., Handler, A., Wall, S. N., & Dwyer, L. (2000). Low-income African-American mothers' perceptions of exposure to racial discrimination and infant birth weight. *Epidemiology, 11,* 337-9.

Krieger, N., & Sidney, S. (1997). Prevalence and health implications of anti-gay discrimination: A study of black and white women and men in the CARDIA cohort. *International Journal of Health Services, 27,* 157-176.

Krieger, N,. Sidney, S., & Coakley, E. (1998). Racial discrimination and skin color in CARDIA: Implications for public health research. *American Journal of Public Health, 88,* 1308-1313.

Ren, X. S., Amick, B. C., & Williams, D. R. (1999). Racial/ethnic disparities in health: The interplay between discrimination and socioeconomic status. *Ethnicity Disease, 9,* 151-165.

Watson, J. M., et al. (2002). Race, socioeconomic status, and perceived discrimination among healthy women. *Journal of Women's Health & Gender-Based Medicine, 11*(5), 441-451.

Yen, I. H., Ragland, D., Breiner, B. A., & Fisher, J. A. (1999). Racial discrimination and alcohol-related behavior in urban transit operators: Findings from the San Francisco Municipal Health and Safety Study. *Public Health Report, 114,* 448-458.

For further discussion, see:

Krieger, N. (2000). Discrimination and health. In L. Berkman & I. Kawachi (Eds). *Social Epidemiology* (pp. 36-75). Oxford: Oxford University Press.

Contact Information

The instrument is available from Nancy Krieger at no cost, under the stipulation that it is cited using both the 1996 CARDIA study and the 1990 article in which the questions were first used (details provided in Dr. Krieger's standard cover letter for the instrument).

Nancy Krieger, Ph.D.
Department of Health and Social Behavior
Harvard School of Public Health
677 Huntington Avenue
Boston, MA 02115, USA

Tel: 617-432-1571 - work

e-mail: nkrieger@hsph.harvard.edu

Race, Racism, Ethnicity, Racial Discrimination, & Related Measures

TITLE OF MEASURE	SCHEDULE OF RACIST EVENTS (SRE)
Source/Primary reference	Landrine, H., & Klonoff, E. A. (1996). The schedule of racist events. *Journal of Black Psychology, 22*, 144-168.
Construct measured	Experiences of specific instances of racial discrimination and racist events
Brief description	The SRE is a self-report inventory containing 18 items that are each rated in three different ways. They are answered once for the frequency in the last year, another time for the frequency in the respondent's lifetime, and a third time for appraising the stressfulness of each event. Responses range from 1 = the event never happened to me, to 6 = the event happens all of the time, for the first two subscales, and 1 = not at all stressful to 6 = very stressful, for the third subscale.
Sample items	▪ How many times have you been treated unfairly by your employers, bosses and supervisors because you are black? ▪ How many times have you been treated unfairly by your coworkers, fellow students and colleagues because you are black?
Appropriate for whom (i.e. which population/s)	African Americans (can be adapted for other minority populations)
Translations & cultural adaptations available	Similar in format and conceptualization to the Schedule of Sexist Events (See entry for Klonoff & Landrine, 1995.)
How developed	The items were written by the authors based on the literature on racism. They conceptualize racist events as analogous to the generic life events and hassles as assessed by popular measures of stressful events. Also, they view racist events as culture-specific, negative life events (i.e., culturally specific stressors). Thus, they modeled their scale after other major general measures of the frequency and appraisal of stressful events.
Psychometric properties	*STUDY SAMPLE*

Participants		Demographics
Sample Size Description		n = 153 Students, faculty, & staff of large university
Age	Range Mean (SD)	15-70 30.14 (11.66)
Gender	Female Male Missing	n = 83 n = 66 n = 4
Race/Ethnicity	African American	100%
Annual Income	Range Mean (SD)	$0 – $80,000 $21,451 ($17,175)
Marital Status	Married Single	n = 40 n = 85

Race, Racism, Ethnicity, Racial Discrimination & Related Measures

TITLE OF MEASURE — SCHEDULE OF RACIST EVENTS (SRE)

<u>VALIDITY</u>
Concurrent Validity

The authors examined the relationships between the scores of the SRE, and the African American Acculturation Scale (AAAS; Landrine & Klonoff, 1994). Mean scores on the SRE subscales are presented below according to AAAS cluster (traditional or acculturated).

SRE Subscale	AAAS Traditional (n = 61)	AAAS Acculturated (n = 75)
Recent Racist Events	46.32	38.67
Lifetime Racist Events	60.62	46.86
Appraised Racist Events	57.59	46.79

<u>RELIABILITY</u>

Subscale	Cronbach's α	Split-half reliability
Recent Racist Events	.95	.93
Lifetime Racist Events	.95	.91
Appraised Racist Events	.94	.92

Comments

- Each of the SRE subscales was higher on average in participants with high stress-related symptoms as measured by the Hopkins Symptom Checklist (HSCL-58; Derogatis, Lipman, Rickles, Uhlenhuth, & Covi, 1974); and higher among cigarette smokers, considered a stress-related behavior.

Subscale	HSCL High (n = 53)	HSCL Low (n = 53)	Nonsmokers (n = 113)	Smokers (n = 24)
Recent Racist Events	46.73	37.95	41.23	44.66
Lifetime Racist Events	59.17	46.84	50.53	62.61
Appraised Racist Events	58.62	43.83	49.42	61.53

- There is also evidence of a relationship between SRE and mental health among African Americans (Klonoff, Landrine, & Ullman, 1999).

- Huebner (2002) adapted this scale to measure discrimination against gay and bisexual men (alpha = .92) and found scores correlated with both physical and mental health outcomes.

Bibliography (studies that have used the measure)

Klonoff, E. A., Landrine, H., & Ullman, J. B. (1999). Racial discrimination and psychiatric symptoms among blacks. *Cultural Diversity and Ethnic Minority Psychology*, 5(4), 329-339.

Race, Racism, Ethnicity, Racial Discrimination, & Related Measures

TITLE OF MEASURE SCHEDULE OF RACIST EVENTS (SRE)

Klonoff, E., & Landrine, H. (1999). Cross-validation of the schedule of racist events. *Journal of Black Psychology*, *25*(2), 231-254.

Contact Information

Elizabeth A. Klonoff
Department of Psychology
San Diego State University
5500 Campanile Drive
San Diego, CA 92182-4611, USA

Race, Racism, Ethnicity, Racial Discrimination & Related Measures

TITLE OF MEASURE	**MODERN RACISM SCALE**
Source/Primary reference	McConahay, J. B., Hardee, B. B., & Batts, V. (1981). Has racism declined in America? It depends upon who is asking and what is asked. *Journal of Conflict Resolution, 25,* 563-579. McConahay, J. B. (1986). Modern racism, ambivalence, and the modern racism scale. In J. Dovidio & S. Gaertner (Eds.) *Prejudice, discrimination and racism* (pp. 91-125). San Diego: Academic Press.
Construct measured	Racial attitudes toward blacks based on four tenets: 1) discrimination is a thing of the past, 2) blacks are pushing too hard, too fast, 3) these tactics are unfair, 4) thus recent gains are undeserved.
Brief description	This measure includes 14 items along two dimensions: 1. Old-Fashioned Racism (7 items) 2. Modern Racism (7 items that ask respondents to what extent they agree or disagree with a set of beliefs that follow the four tenets outlined above)
Sample items	Old-Fashioned Racism: - It is a bad idea for blacks and whites to marry one another. - Black people are generally not as smart as whites. Modern Racism: - Over the past few years, blacks have gotten more economically than they deserve. (agree-disagree) - It is easy to understand the anger of black people in America. (disagree-agree) - Over the past few years, the government and news media have shown more respect for blacks than they deserve. (agree-disagree) - How many black people in XX County do you think miss out on good housing because white owners won't rent or sell to them? (from many to none)
Appropriate for whom (i.e. which population/s)	Adolescents or adults
Translations & cultural adaptations available	None known

Race, Racism, Ethnicity, Racial Discrimination, & Related Measures

TITLE OF MEASURE	MODERN RACISM SCALE
How developed	The authors began with the Old-Fashioned Racism (OFR) Scale but found that the items were so reactive that they pulled for socially desirable responses and were so blatant that some study participants refused to answer them. By 1976, there had been enough experience with these items to formulate a general definition of *symbolic racism* or *modern racism* to include components of racial attitudes missed by the OFR scale. A new set of items was generated from this definition. The first version of the MRS was used with adult community residents (Studies 1 and 2 below). Another version of the scale was used in several college student samples (Study 3). Over the years, the scale has been further refined.
Psychometric properties	*STUDY SAMPLES*

Participants	Study 1	Study 2	Study 3
Sample Size	$n = 879$	$n = 709$	$n = 167$
Description	White adults (18 years and older) residing in Louisville and Jefferson County, Kentucky, 1976	White adults (18 years and older) residing in Louisville and Jefferson County, Kentucky, 1977	White undergraduate students, enrolled in introductory psychology classes at Duke University, 1984
Gender	Not reported	Not reported	Not reported

VALIDITY

Construct Validity

A number of factor analyses were performed on various combinations of Modern and Old-Fashioned Racism Scale items. Across analyses, the Modern Racism items loaded most highly on one factor, while the Old-Fashioned items loaded on another factor. These results support the notion that Modern Racism is distinct from Old-Fashioned Racism, although correlated ($r = .68, .70, .59$ in the three study samples, respectively).

The Modern Racism Scale correlated with strength of opposition to busing in Louisville in surveys done during the conflict there in 1976 ($r = .511$) and 1977 ($r = .391$).

The scale also correlated significantly with voting preferences for a black candidate versus a white incumbent for mayor of Los Angeles in both 1969 and 1973 (McConahay & Hough, 1976). Those whites scoring high on the scale were more likely than those with low scores to vote for the white candidate in 1969 ($r = .365$) and 1973 ($r = .338$), and these correlations were still significant after controlling for political

Race, Racism, Ethnicity, Racial Discrimination & Related Measures

TITLE OF MEASURE	MODERN RACISM SCALE

conservatism (partial $r = .309$ and $.300$, respectively). All correlations were statistically significant.

Concurrent Validity

The Modern Racism Scale correlated with several other scales designed to assess related constructs.

Scales	Sample	$r =$
Sympathetic Identification with the Underdog (Schuman & Harding, 1963)	Louisville adults	-.299
Antiblack Feeling measured by the Feeling Thermometer (Campbell, 1971)	Louisville adults	.383
Feeling Thermometer	College students over 16 years	Average $r = .441$

Scores on the Modern Racism Scale did not correlate with the Just World Scale in repeated college student samples. Because the Feeling Thermometer and the Old-Fashioned Racism Scales are accepted as face-valid measures of racism and the belief in a just world has been proposed as an alternative explanation for high scores on the moralistic items in the scale, this is strong evidence for the concurrent/criterion validity of the Modern Racism Scale.

The strongest evidence for the validity of the Modern Racism Scale emerged from an experimental study of simulated hiring decisions using white college student participants, in which MRS scores were related to evaluations of the black candidates (McConahay, 1983).

RELIABILITY

Internal Consistency

Scale	Study 1 $\alpha =$	Study 2 $\alpha =$	Study 3 $\alpha =$
Modern Racism	.75	.79	Range: .81 - .86

Test-Retest Reliability

Ranges from .72 to .93 across a number of samples.

Comments	▪ This scale assesses a component of racist attitudes that is particularly relevant to work situations in that it gets at assessments of and reactions to progress in the recent past.

Race, Racism, Ethnicity, Racial Discrimination, & Related Measures

TITLE OF MEASURE	MODERN RACISM SCALE
	<ul><li>It has been shown to be related to work behaviors in hiring simulations.</li><li>Given the hypotheses, the study samples were 100% white by design. However, it would be useful to assess the scale's validity and reliability for multiple ethnic/racial groups.</li><li>Gender is not reported and thus applicability of the scale for women is unknown.</li></ul>
Bibliography (studies that have used the measure)	McConahay, J. B. (1982). Self-interest versus racial attitudes as correlates of anti-busing attitudes on Louisville: Is it the buses or the blacks? *Journal of Politics, 44,* 692-720. McConahay, J. B. (1983). Modern racism and modern discrimination: The effects of race, racial attitudes, and context on simulated hiring decisions. *Personality and Social Psychology Bulletin, 9*(4), 551-558.
Contact Information	John B. McConahay Public Policy Studies Box 90245 Duke University Durham, NC 27706, USA Tel: 919-613-7324 e-mail: mcconaha@pps.duke.edu

Race, Racism, Ethnicity, Racial Discrimination & Related Measures

TITLE OF MEASURE	PERCEIVED RACISM SCALE
Source/Primary reference	McNeilly, M. D., Anderson, N. B., Armstead, C. A., Clark, R., Corbett, M., Robinson, E. L., Pieper, C. F., & Lepisto, E. M., (1996). The Perceived Racism Scale: A multidimensional assessment of the experience of white racism among African Americans. *Health, Ethnicity and Disease, 6,* 154-166
Construct measured	Perceived exposure to racism
Brief description	PRS is a 51-item instrument. The first section has 43 items and asks the respondents to rate the frequency with which they have been exposed to racist events in four domains: job, academic, public, and racist statements (0 = not applicable, 7 = several times a day). The second section includes 8 items, which require respondents to indicate the emotional appraisal of each event (e.g., angry, frustrated, sad, powerless, etc.). Section three requires respondents to indicate coping strategies that have been used for each event (e.g., speaking up, ignoring it, etc.).
Sample items	■ Because I am black, I am assigned to the jobs no one else wants. ■ I have been made to feel uncomfortable in the classroom. ■ I have been refused housing because I am black. ■ When I go shopping, I am often followed.
Appropriate for whom (i.e. which population/s)	African-American adults (can be adapted for other minority populations)
Translations & cultural adaptations available	None known
How developed	Items for the scale were empirically derived by collecting data from 165 African American psychology students at North Carolina Central University (108 females, 57 males) and 25 individuals from the community (15 females, 10 males). The age range of the participants was 18-46 (M = 21, SD = 4.8). They were asked to list their personal experiences of racism and the feelings related to these experiences. Their responses were then categorized into four domains: 1) on the job; 2) in academic settings; 3) in the public realm; 4) exposure to racist statements. The items most frequently mentioned were selected for the scale. The new instrument was piloted with 10 students and 10 individuals from

Race, Racism, Ethnicity, Racial Discrimination, & Related Measures

TITLE OF MEASURE: **PERCEIVED RACISM SCALE**

the community, who provided feedback on content, wording, response format, and instructions.

Psychometric properties

STUDY SAMPLES

Participants		Student Sample 1	Community Sample	Student Sample 2	Student Sample 3
Sample Size		n = 110	n = 104	n = 59	n = 32
Age	Range	18-35	18-73	18-39	-
	M (SD)	21.2 (2.9)	33.7 (12.48)	21.6 (4.17)	21.6 (3.5)
Gender	Female	n = 73	n = 84	n = 41	n = 28
	Male	n = 37	n = 20	n = 18	n = 4

VALIDITY

Construct Validity

Exploratory principal component factor analyses were performed using both orthogonal and oblique rotations. The items were divided according to their type: frequency of exposure (43 questions) and emotional and coping responses (8 questions). The samples that were used in these analyses were student samples 1 and 2 and the community sample.

Both orthogonal and oblique rotations resulted in very similar factors. Factor rotations for over the past year, for over one's lifetime, and for the frequency of exposure were nearly identical. Five factors emerged for the exposure items (racism on the job, racism in academic settings, overt racism in public, subtle racism in public, and racist statements), and nine factors for the emotional and behavioral coping (anger/frustration, depressed affect, feeling strengthened, trying to change things, avoiding/ignoring, praying, forgetting it, getting violent, and speaking up).

RELIABILITY

Internal Consistency

Internal consistency was assessed based on the responses from student samples 1 and 2 plus the community sample ($n = 273$):

Subscale	Scale $\alpha =$	α for the individual factors
Frequency of Exposure Domains (items 1-43)	.96	.84-.93
Emotional and Behavioral Coping Responses (items 44-51)	.94	.64-.95

Race, Racism, Ethnicity, Racial Discrimination & Related Measures

TITLE OF MEASURE — PERCEIVED RACISM SCALE

Test-Retest Reliability

Student sample 2 was tested over an interval of two weeks. The researchers asked student sample 3 to think of a racist event that happened to them in each domain and to complete the emotional and coping subscales with the incidents in mind. They were asked to recall the same incidents when completing the scale two weeks later.

Subscale	Inter-class Correlations	
	Student Sample 2	*Student Sample 3*
Frequency of Exposure	.70-.80	-
Emotional Responses	.50-.78	.43-.87
Coping Responses	.59	.60

Comments
- Includes items related to discrimination at work

Bibliography (studies that have used the measure)

Contact Information

Maya Dominguez McNeilly
Box 3003
Duke University Medical Center
Durham, NC 27710, USA

e-mail: maya@geri.duke.edu

Race, Racism, Ethnicity, Racial Discrimination, & Related Measures

TITLE OF MEASURE	MOTIVATION TO RESPOND WITHOUT PREJUDICE
Source/Primary reference	Plant, E. A., & Devine, P. G. (1998). Internal and external motivation to respond without prejudice. *Journal of Personality and Social Psychology, 75*(3), 811-832.
Construct measured	Sources of internal and external motivations to respond without prejudice toward blacks
Brief description	The final scale consists of 10 items, rated on a 9-point scale from 1 = strongly disagree to 9 = strongly agree. There are two subscales: 1. Internal Motivation to Respond Without Prejudice (IMS), with 5 items 2. External Motivation to Respond Without Prejudice (EMS), with 5 items
Sample items	The IMS subscale: - I attempt to act in non-prejudiced ways towards black people because it is personally important to me. - Being non-prejudiced towards black people is important to my self-concept. The EMS subscale: - I attempt to appear non-prejudiced towards black people to avoid disapproval from others. - I try to act non-prejudiced toward black people because of pressure from others.
Appropriate for whom (i.e. which population/s)	White or non-black adults
Translations & cultural adaptations available	Similar scales adapted from the original have been used to measure motivation to respond without sexism, prejudice toward fat people, and prejudice toward homosexuals.
How developed	Phase 1: In the first phase, an initial 19-item questionnaire was created by the authors. Phase 2: The final scales were developed using exploratory and confirmatory factor analyses. Two factors – the IMS and EMS subscales

Race, Racism, Ethnicity, Racial Discrimination & Related Measures

TITLE OF MEASURE — MOTIVATION TO RESPOND WITHOUT PREJUDICE

- emerged. The discriminant and convergent validity of the IMS and EMS were examined by comparing them to other measures.

Phase 3: The final phase involved demonstrating the predictive validity of the IMS and EMS by examining (i) people's affective reactions to living up to own-based (internal) and other-based (external) standards for how blacks should be treated, and (ii) the extent to which people reported endorsing the stereotype of blacks under either private and anonymous or public conditions.

Psychometric properties

STUDY SAMPLES:

Participants		Sample 1	Sample 2	Sample 3
Sample Size		n = 135	n = 245	n = 1,363
Description		Introductory psychology students	Introductory psychology students	Introductory psychology students
Gender	Females	78%	74%	60%
	Males	22%	26%	40%
Ethnicity	Whites	94%	84%	85%
	Non-whites	6%	16%	15%

Samples 1 & 2: The first two samples completed the initial 19-item questionnaire in medium-sized groups and received an extra course credit for their participation.

Sample 3: The third sample completed the final set of 10 items (refined questionnaire), and received an extra course credit for their participation. A sub-sample of Sample 3 filled out the IMS and EMS scales 9 weeks after the mass testing session to examine the test-retest reliabilities.

VALIDITY

Construct Validity

An exploratory factor analysis for Sample 1 revealed that there were two strong factors and two weak factors with eigenvalues over 1.00. The first factor accounted for 28% of the variance (eigenvalue 5.33) and consisted of items about internal motivation to respond without prejudice. The second factor accounted for 20% of the variance (eigenvalue 3.74) and included items that assessed external motivation to respond without prejudice. Four items were dropped because they either (i) loaded on both factors, possibly not differentiating internal from external motivation to respond without prejudice, or (ii) failed to load on either of the factors with a loading of .50 or above.

Race, Racism, Ethnicity, Racial Discrimination, & Related Measures

TITLE OF MEASURE	MOTIVATION TO RESPOND WITHOUT PREJUDICE

Confirmatory factor analysis across all three samples revealed that the two-factor model provided a significantly better fit of data than the one-factor model.

Concurrent Validity

Correlations between the IMS and EMS as well as other measures

Measure	IMS	EMS
Motivation measures		
• IMS	-	-.15*
• EMS	-.15*	-
Prejudice measures		
• Modern Racism Scale (McConahay et al. 1981)	-.57**	-.22**
• Pro-black Scale (Katz & Hass, 1988)	.24**	.03
• Anti-black Scale (Katz & Hass, 1988)	-.48**	.12
• Attitude Toward blacks Scale (Brigham, 1993)	.79**	-.27**
• Right-Wing Authoritarianism Scale (Altemeyer, 1981)	-.24**	.13*
• Protestant Work Ethic Scale (Katz & Hass, 1988)	-.18*	.12
• Humanitarianism-Egalitarianism Scale (Katz & Hass, 1988)	.45**	-.09
Social evaluation and self-perception measures		
• Fear of Negative Evaluation Questionnaire (Leary, 1983a, & Watson & Friend, 1969)	.11	.14*
• Interaction Anxiousness Scale (Leary, 1983b)	-.03	.16*
• Marlowe-Crowne Social Desirability Scale (Crowne & Marlowe, 1960)	-.07	-.11
• Self-Monitoring Scale (Snyder & Gangestad, 1986)	-.02	-.01

N = 247
*p < .05; **p < .01

Correlations between the IMS, the EMS, the Attitude Toward blacks Scale (ATS), and the Motivation to Control Prejudiced Reactions Scale (MCPR)

Measure	IMS	EMS	ATS
Motivation to Control Prejudiced Reactions Scale (MCPR; Dunton & Fazio, 1997)	.22*	.36**	.20*
• Concern with acting prejudiced	.38**	.26*	.35**
• Restraint to avoid dispute	-.21**	.35**	-.20*
Attitude Toward blacks Scale (ATS) (Brigham, 1993)	.72**	-.33**	-

N = 119
*p < .05; **p < .01

Race, Racism, Ethnicity, Racial Discrimination & Related Measures

TITLE OF MEASURE — MOTIVATION TO RESPOND WITHOUT PREJUDICE

RELIABILITY

Internal Consistency and Test-Retest Reliability

Cronbach α reliability coefficients of the IMS and EMS across all three samples, as well as the IMS and EMS test-retest correlation coefficients

Subscales	Sample 1 (n = 135) α =	Sample 2 (n = 245) α =	Sample 3 (n = 1,352) α =	Test-retest reliability (sub-sample of Sample 3) (n = 159) r =
IMS	.85	.84	.81	.77
EMS	.79	.76	.80	.60

Comments

- The scales measure mostly independent constructs and have good convergent and discriminant validity.

- The different studies of this measure support the argument that there are distinct internal and external motivations underlying people's desire to avoid prejudiced responses.

- Correlations of the IMS and EMS with measures of racial attitudes suggest that traditional attitude measures are more strongly related to internal than external motivation to respond without prejudice.

- Although the EMS subscale seems to be somewhat related to traditional measures of prejudice and social anxiety, it appears to measure something beyond social anxiety.

- During the predictive validation study of Phase 3, where the participants were asked to report the extent to which they endorsed stereotypes of Blacks, the experimenter was an advanced student at the University who was likely to be perceived as a representative of the campus and its well-understood non-prejudiced standards. When reporting responses directly to this person, it is possible that the respondents would be more likely to comply with normative expectations and, thus, avoid prejudiced responses.

Bibliography (studies that have used the measure)

Amodio, D. M., Harmon-Jones, E., & Devine, P. G. (2003). Individual differences in the activation and control of affective race bias as assessed by startle eyeblink responses and self-report. *Journal of Personality and Social Psychology, 84,* 738–753.

Devine, P. G., Plant, E. A., Amodio, A. M., Harmon-Jones, E., & Vance, S. L. (2002). Exploring the relationship between implicit and explicit

Race, Racism, Ethnicity, Racial Discrimination, & Related Measures

TITLE OF MEASURE	MOTIVATION TO RESPOND WITHOUT PREJUDICE

prejudice: The role of motivations to respond without prejudice. *Journal of Personality and Social Psychology, 82,* 835-848.

Plant, E. A. (2004). Responses to interracial interactions over time. *Personality and Social Psychology Bulletin, 30,* 1458-1471.

Plant, E. A., & Devine, P. G. (2001). Responses to other-imposed pro-black pressure: Acceptance or backlash? *Journal of Experimental Social Psychology, 37,* 486–501.

Plant, E. A., Devine, P. G., & Brazy, P. C. (2003). The bogus pipeline and motivations to respond without prejudice: Revisiting the fading and faking of prejudice. *Group Processes and Intergroup Relations, 6,* 187-200.

Contact Information

E. Ashby Plant
Department of Psychology
102d Psychology Building
Florida State University
Tallahassee, FL 32306-1270, USA

Tel: 850-644-5533

e-mail: plant@psy.fsu.edu

www.psy.fsu.edu/faculty/plant.dp.html

Race, Racism, Ethnicity, Racial Discrimination & Related Measures

TITLE OF MEASURE	**ACCULTURATIVE STRESS SCALE (ACS)**
Source/Primary reference	Salgado de Snyder, V. N. (1987). Factors associated with acculturative stress and depressive symptomatology among married Mexican immigrant women. *Psychology of Women Quarterly, 11*, 475-488.
Construct measured	Stress associated with acculturation
Brief description	The ACS scale is a 12-item measure which assesses stressors in the familial, marital, social, financial, and environmental domains. For each item, the respondent is asked whether she has experienced the potential stressful situation in the last three months. If the answer is affirmative, people are asked to further respond on a 4-point Likert-type scale to indicate the degree of stressfulness in each situation (0 = not stressful at all to 4 = very stressful). A high score indicates high stress.
Sample items	<ul><li>Not having enough money to pay debts.</li><li>Not being able to communicate in English.</li><li>Being discriminated against because of being Mexican.</li><li>Having accented speech in English.</li></ul>
Appropriate for whom (i.e. which population/s)	Adult Spanish-speaking immigrant women
Translations & cultural adaptations available	Spanish translation
How developed	The ACS items were derived from the original 172-item Latin American Stress-Inventory (LAS-I) developed by a research group of the Spanish Speaking Mental Health Research Center (Cervantes, Padilla, & Salgado de Snyder, 1987).

Race, Racism, Ethnicity, Racial Discrimination, & Related Measures

TITLE OF MEASURE	ACCULTURATIVE STRESS SCALE (ACS)

Psychometric properties

<u>STUDY SAMPLES</u>

Participants		Demographics
Sample Size		n = 140
Description Selected from the files of 1984-1985 marriage licenses of the County of Los Angeles.		Married Mexican immigrant women, who are married for the first time and not born earlier than 1950.
Age	Range	17-49
	Mean	25.7
Children	Children ranging from 2 months to 19 years of age	50%
	No children	50%
Religion	Catholic	87.1%
	Protestants, Baptists, and Jehovah's Witness	12.8%
Language skills	Fluent in spoken English	21.4%
	Speaking knowledge of English	57.8%
	Only Spanish and no English	20%
Education	Range	0-20 years
	Mean	9.4
Employment status	Housewives	50%
	Employment outside homes	50%
	- skilled	33%
	- semi-skilled	59%

<u>VALIDITY</u>

Concurrent Validity

A significant correlation between acculturative stress and depressive symptomatology was observed $r = .40, p < .001$.

<u>RELIABILITY</u>

Internal Consistency

Cronbach's α coefficient of the ACS scale was 0.65.

Comments

- The measure addresses stresses in multiple domains and could be adapted to be more specific to the work setting.

- There were problems locating potential participants and the response rate was 21.5%. Due to the limitations of the sampling criteria and a self-selection bias, the results of the study must be interpreted with caution.

Race, Racism, Ethnicity, Racial Discrimination & Related Measures

TITLE OF MEASURE	ACCULTURATIVE STRESS SCALE (ACS)
	▪ The author notes that a strict random sampling procedure is not possible when doing research with undocumented immigrants because of their clandestine status and fears about the consequences of participating in a study.
Bibliography (studies that have used the measure)	
Contact Information	Dra. V. Nelly Salgado de Snyder Directora de Salud Comunitaria y Bienestar Social Investigadora en Ciencias Medicas "F" Centro de Investigación en Sistemas de Salud Instituto Nacional de Salud Pública, Mexico Tel: +52-777- 329-3019 Fax: +52-777-311-1156 e-mail: nsnyder@insp.mx www.insp.mx

Race, Racism, Ethnicity, Racial Discrimination, & Related Measures

TITLE OF MEASURE	CULTURAL MISTRUST INVENTORY (CMI)
Source/Primary reference	Terrell, F., & Terrell, S. (1981). An inventory to measure cultural mistrust among blacks. *The Western Journal of Black Studies, 5*(3), 180-185.
Construct measured	Beliefs about the extent to which African Americans should trust Euro-Americans
Brief description	This instrument consists of 48 items, divided into subscales that measure mistrust of blacks toward whites in four different domains: 1. Political and legal system 2. Work and business interactions 3. Education and training 4. Interpersonal and social contexts All items rated on a 9-point scale from 1 = not in the least agree to 9 = entirely agree.
Sample items	- Whites are usually fair to all people regardless of race. (work/business) - Black students can talk to white teachers in confidence without fear that the teacher will use it against him or her later. (education) - Blacks should be suspicious of a white person who tries to be friendly. (interpersonal) - White politicians will promise blacks a lot but deliver little. (political)
Appropriate for whom (i.e. which population/s)	African American adults (can be adapted for other minority populations)
Translations & cultural adaptations available	None known
How developed	The authors reviewed the literature to develop items covering each of four domains: 1) Political and legal system, 2) Work and business interactions, 3) Education and training, and 4) Interpersonal and social contexts. Four black psychologists independently rated each item for clarity and domain appropriateness. The items that were considered unclear or inappropriate were rewritten or eliminated. This process continued until all judges agreed on the 81 items that composed the initial Cultural

Race, Racism, Ethnicity, Racial Discrimination & Related Measures

Title of measure	**Cultural Mistrust Inventory (CMI)**
	Mistrust Inventory. Then 23 items were eliminated based on their high correlation with the Social Desirability Scale (Jackson, 1970). An item discrimination analysis led to elimination of 9 additional items that were endorsed by most respondents. Finally, one item was eliminated because it correlated more highly with another subscale than its own.
Psychometric properties	**STUDY SAMPLE** <table><tr><th>Participants</th><th>Demographics</th></tr><tr><td>Sample Size</td><td>$n = 172$</td></tr><tr><td>Description</td><td>African American first- and second-year male college students</td></tr></table> **VALIDITY** *Construct Validity* An F-test was computed between the Racial Discrimination Index (Terrell & Miller, 1980) quartile groups and scores on the CMI, to test the hypothesis that being a victim of racial discrimination would be associated with scores on the CMI; $F = 14.01$ ($p < .001$). Inter-scale correlation coefficients were low (ranging from 0.11 to 0.23), supporting the notion of four separate domains. **RELIABILITY** *Internal Consistency* Internal reliability was assessed by computing Pearson item-total scale score correlations; all items had statistically significant correlations ($p = 0.05$). *Test-Retest Reliability* Test-retest reliability was measured over a two-week interval ($n = 69$) with a result of 0.86 (statistic not specified).
Comments	- Need for further research, including a factor analysis of the domains of this inventory. - Although the measure is somewhat old (1980), most items still seem relevant today. - The study samples were 100% male. It would be important to assess applicability and norms for women.

Race, Racism, Ethnicity, Racial Discrimination, & Related Measures

TITLE OF MEASURE	**CULTURAL MISTRUST INVENTORY (CMI)**
Bibliography (studies that have used the measure)	Thompson, C. E., Neville, H., Weathers, P. L., Poston, W. C., & Atkinson, D. R. (1990). Cultural mistrust and racism reaction among African American students. *Journal of College Student Development, 31,* 162-168.
Contact Information	Francis Terrell University of North Texas Department of Psychology P.O. Box 311280 Denton, TX 76203-1280, USA Tel: 940-565-2671 e-mail: terrellf@unt.edu

Race, Racism, Ethnicity, Racial Discrimination & Related Measures

TITLE OF MEASURE	**RACISM REACTION SCALE (RRS)**
Source/Primary reference	Thompson, C. E., Neville, H., Weathers, P. L., Poston, W. C., & Atkinson, D. R. (1990). Cultural mistrust and racism reaction among African American students. *Journal of College Student Development, 31,* 162-168.
Construct measured	Sense of being differentially treated
Brief description	The inventory includes 6 statements related to a sense of being personally threatened, differentially treated, or singled out for differential treatment. Each item is rated on a 7-point scale from 1 = strongly agree to 7 = strongly disagree.
Sample items	▪ I have to be prepared to deal with a threatening environment. ▪ Other students are surprised to learn that I have some of the same feelings and goals that they have.
Appropriate for whom (i.e. which population/s)	Students (can be adapted for use in a work setting)
Translations & cultural adaptations available	None known
How developed	Initially, 19 items were chosen from the statements of racism reactions made by African American students who participated in a racial awareness program at a predominantly white university. The statements were reworded to conceal references to race.
Psychometric properties	*STUDY SAMPLE*

Participants		Demographics	
Sample Size & Description		African American *n* = 87	Euro-American *n* = 70
Gender	*Female*	*n* = 49	*n* = 39
	Male	*n* = 37	*n* = 31
Age	Range	17-42	
	Mean (SD)	21 (4.2)	
Academic Level	Freshman	*n* = 35	
	Sophomore	*n* = 41	
	Junior	*n* = 38	
	Senior	*n* = 43	

VALIDITY

Construct Validity

The scores were compared between African American and Euro-American students, using t-test for independent means. Scores were higher in the former group for 16 of 19 items. Six differences had a

Race, Racism, Ethnicity, Racial Discrimination, & Related Measures

TITLE OF MEASURE — RACISM REACTION SCALE (RRS)

statistical significance exceeding a .05 alpha level and were chosen for inclusion in RRS.

Questions	African American: Mean (SD)	Euro-American: Mean (SD)
People keep asking me about my manner of grooming.	2.7 (1.9)	2.1 (1.8)
I have to be prepared to deal with a threatening environment.	4.6 (2.1)	3.3 (1.9)
Other students are surprised to learn that I have some of the same feelings and goals that they have.	3.6 (2.0)	2.8 (1.7)
When I walk into class, everyone turns his or her head to look at me.	3.5 (2.1)	2.4 (1.4)
Professors don't expect me to perform as well as other students.	2.2 (1.6)	1.4 (1.0)
The other students expect me to do poorly in our classes.	2.0 (1.5)	1.6 (1.2)

Concurrent Validity

Pearson correlations were calculated between the RRS and the 3 subscales of the Cultural Mistrust Inventory (CMI) (Terrell & Terrell, 1981).

	CMI Subscale		
	Interpersonal Relations	Education & Training	Combined
Scale	r =	r =	r =
RRS	.22	.43	.34

Comments
- Although the scale was developed for use with students, it could be adapted for use with a broader adult working population.

Bibliography (studies that have used the measure)

Contact Information

Chalmer E. Thompson
Department of Counseling & Educational Psychology
Indiana University
201 N Rose Ave., Room 4054
Bloomington, IN 47405, USA

Tel: 812-856-8319

e-mail: chathomp@indiana.edu

Race, Racism, Ethnicity, Racial Discrimination & Related Measures

TITLE OF MEASURE	**INDEX OF RACE-RELATED STRESS (IRRS)**
Source/Primary reference	Utsey, S. O., & Ponterotto, J. G. (1996). Development and validation of the Index of Race-Related Stress (IRRS). *Journal of Counseling Psychology, 43*(4), 490-501.
Construct measured	Stress associated with specific events of racism and discrimination
Brief description	The instrument is a 46-item self-report measure of the stress experienced by African Americans as a result of daily racism and discrimination. The scale is a multidimensional measure (consisting of 4 subscales and a Global Racism measure) that takes into consideration both frequency and appraisal. 1. Cultural Racism Subscale - 16 items intended to measure the experience of racism when one's culture is denigrated 2. Institutional Racism subscale -11 items to assess the experience of racism embedded in institutional policies 3. Individual Racism subscale - 11 items to assess the experience of racism on the interpersonal level 4. Collective Racism subscale - 8 items to assess racism experienced as the concerted efforts of whites/non-blacks to restrict African Americans' rights Respondents are asked to indicate which of the listed events they (or their family members) have experienced in their lifetimes. Then the chosen events are assessed on a 5-point rating scale ranging from 0 = never happened to 4 = event happened and I was extremely upset. Ratings on items are summed for total IRRS score. The Global Racism score is derived by weighting each of the subscales and then summing.
Sample items	- While shopping at the store, the sales clerk assumed that you couldn't afford certain items (i.e., you were directed toward items on sale). (Individual racism) - You have attempted to hail a cab, but they refused to stop; you think because you are black. (Collective racism) - You seldom hear or read anything positive about black people on radio, TV, newspapers or in history books. (Cultural racism) - You did not get the job you applied for although you were well qualified; you suspect because you are black. (Institutional racism)

Race, Racism, Ethnicity, Racial Discrimination, & Related Measures

TITLE OF MEASURE	INDEX OF RACE-RELATED STRESS (IRRS)
Appropriate for whom (i.e. which population/s)	African-American adults (can be adapted for other minority populations)
Translations & cultural adaptations available	None known
How developed	The initial items were developed based on interviews with male and female African Americans from various backgrounds, literature review, and the personal life experience of the first investigator (an African American male). A total of 74 items reflecting experiences of racism and discrimination were generated, then placed on a 5-point scale. The scale range was 1 = no reaction to 5 = rage. Respondents had to rate only the events they had experienced. Further analysis of two population samples (described below) yielded a final version of the scale with 46 items.
Psychometric properties	*See below*

STUDY SAMPLES

Participants		Study 1	Study 2		Study 3	
Sample		Overall	Overall	Subsample	Sample 1	Sample 2
Sample Size		n = 302	n = 310	n = 31	n = 31	n = 19
Description		African Americans	African Americans	23 whites 8 Asians	African American college students	African Americans from an adult education program
Age	Range	18-61	17-76	17-76	-	-
	M (SD)	26.77 (9.02)	23.38 (7.74)	23.38 (3.79)	20.48 (3.78)	29.42 (9.42)
Gender	Female	167 (55%)	207 (67%)	16 (55%)	21 (67%)	15 (79%)
	Male	115 (38%)	92 (30%)	15 (45%)	9 (29%)	4 (21%)
	Missing	19 (7%)	11 (3%)		1 (3%)	

VALIDITY

Content Validity

The authors conducted a focus group composed of 5 African Americans to evaluate the content validity of the initial items. As a result, the Likert-type scale was modified to 1 = no reaction to 4 = extremely upset by the event. Some items were rewritten and some omitted.

In the next step, five additional experts judged the domain appropriateness of each item.

A pilot study was conducted throughout the U.S. ($n = 377$: 203 women, 163 men, 11 unknown). This resulted in adding another point (0 = this never happened to me) to the existing Likert scale.

Race, Racism, Ethnicity, Racial Discrimination & Related Measures

TITLE OF MEASURE   INDEX OF RACE-RELATED STRESS (IRRS)

Construct Validity

Pilot Study: Principal components analysis on 67 items showed that up to four components were interpretable.

    Component 1: cultural racism
    Component 2: institutional-level racism
    Component 3: individual-level racism
    Component 4: collective racism (extension of Essed, 1990).

The researchers performed 1-, 2-, 3-, 4-component extractions with both oblique and orthogonal methods. Items with loadings of .35 or higher on a single factor were retained, yielding 59 questions.

Study 1 assessed the principal-component structure of the revised scale. The most interpretable and conceptually supported was the four-component orthogonal solution, which accounted for 38% of the common variance. As a result of these findings, 13 items were eliminated from the scale.

Pearson product-moment correlation coefficients among the subscales of the IRRS were low to moderate, supporting conceptualization of the subscales as distinct measures of the stress experienced by African Americans.

	Subscale		
	2	3	4
Subscale	$r =$	$r =$	$r =$
1 Cultural Racism	.42**	.56**	.30**
2 Institutional Racism		.57**	.58**
3 Individual Racism			.39**
4 Collective Racism			

**$p < .01$

Study 2: A confirmatory factor analysis of the scale component structure was conducted to investigate the construct validity of the scale. Subscale inter-correlation coefficients remained low to moderate, as in Study 1.

Concurrent Validity

Study 2: IRRS scores were compared with a second measure of racism (Racism and Life Experience Scale - RaLES-B, Harrell, 1994) and with a second measure of perceived stress (Perceived Stress Scale – PSS, Cohen, Karmarck, & Mermelstein, 1983). The IRRS subscales and the global (total z-weighted) scores were generally strongly associated with subscales of RaLES-B and the PSS, using Pearson product-moment correlation coefficients:

Race, Racism, Ethnicity, Racial Discrimination, & Related Measures

TITLE OF MEASURE   INDEX OF RACE-RELATED STRESS (IRRS)

IRRS Scale	RaLES-B (n = 57)			PSS (n = 51)
	Self	Group	Global	
Cultural Racism	.04	.46**	.29*	.31*
Institutional Racism	.39**	.36**	.44**	.15
Individual Racism	.23*	.31**	.31**	.24*
Collective Racism	.25*	-.02	.15	.09
Global Racism	.30*	.38**	.39**	.24*

$*p < .05; **p < .01$

IRRS subscale scores were compared between black and 31 non-black (white and Asian) respondents, using multivariate analysis of variance. Blacks scored significantly higher on each IRRS subscale (all p-values < 0.01).

RELIABILITY

Internal Consistency

Internal consistency was high for each IRRS subscale.

Scale	Study 1: Cronbach's α	Study 2: Cronbach's α
Cultural Racism	.87	.89
Institutional Racism	.85	.82
Individual Racism	.84	.84
Collective Racism	.79	.74

Test-Retest Reliability

Test-retest reliability was assessed in Study 3 over a three-week interval for the first sample and a two-week interval for the second sample.

Scale	Sample 1 Reliability Coefficients	Sample 2 Reliability Coefficients
Cultural Racism	.77	.58
Institutional Racism	.69	.71
Individual Racism	.61	.54
Collective Racism	.79	.75

Comments

- Appears to be a reliable and valid measure.
- Addresses the multidimensionality of the experience of race-related stress.

Race, Racism, Ethnicity, Racial Discrimination & Related Measures

TITLE OF MEASURE	INDEX OF RACE-RELATED STRESS (IRRS)

	• The Institutional Racism subscale, which actually appears to assess individuals' experiences of institutional practices, has the items most relevant to workplace issues.
Bibliography (studies that have used the measure)	
Contact Information	Shawn Utsey & Joseph Ponterotto Psychological and Educational Services Fordham University at Lincoln Center New York, NY 10023, USA e-mail: utsey@mary.fordham.edu

Race, Racism, Ethnicity, Racial Discrimination, & Related Measures

TITLE OF MEASURE	RACE-RELATED STRESS
Source/Primary reference	Williams, D. R., Yu, Y., Jackson, J. S., & Anderson, N. B. (1997). Racial differences in physical and mental health. *Journal of Health Psychology, 2*(3), 335-351.
Construct measured	Experiences of lifetime discrimination and everyday discrimination
Brief description	The 12-item instrument includes two sets of questions: 1. Discrimination (3 items, count ranging from "none" to "three or more events") 2. Everyday Discrimination (9 items, rated from "never" to "often") (based on Essed, 1991) Following each section, respondents are asked to rank the three most common reasons for their unfair treatment from a list of nine possible reasons.
Sample items	Lifetime Experiences of Discrimination: Do you think you have ever been unfairly: - not been hired for a job? - fired or denied promotion? - stopped, searched, questioned, physically threatened, or abused by police? Everyday Discrimination: How often: - are you treated with less courtesy than others? - do you receive poorer service than others in restaurants? - do people act as if you are not smart? - are people afraid of you?
Appropriate for whom (i.e. which population/s)	Adults
Translations & cultural adaptations available	None known
How developed	Items were written by the study authors. No additional detail is provided.

Race, Racism, Ethnicity, Racial Discrimination & Related Measures

TITLE OF MEASURE   RACE-RELATED STRESS

Psychometric properties	*STUDY SAMPLE*

Participants	Demographics	
Sample Size	$n = 1,106$	
Description	Adults residing in Wayne, Oakland, and Macomb Counties in Michigan, including the city of Detroit	
	Age Range	**18 years and older**
Gender		Not reported†
Race/Ethnicity‡	Black	$n = 586$
	White	$n = 520$

†While the gender breakdown of the sample was not reported, gender was "controlled for" in regression analyses reported in the article.

‡Although respondents included a total of 33 Asians, Native Americans, and Hispanics, data from these participants were excluded from analyses.

RELIABILITY

Internal Consistency

Subscale	Cronbach's $\alpha =$
Everyday Discrimination Scale	.88

Comments	<ul><li>The Everyday Discrimination scale was associated cross-sectionally with four different indicators of health status and accounted for a large proportion of the differences in health between blacks and Whites, beyond the effect of socioeconomic status (Williams et al., 1997).</li><li>Others have reworded questions to improve clarity (e.g Hughes & Johnson, 2001).</li><li>More psychometric assessment is needed.</li></ul>
Bibliography (studies that have used the measure)	Hughes, D., & Johnson, D. (2001). Correlates in children's experiences of parents' racial socialization behaviors. *Journal of Marriage and Family, 63*(4), 981-996. Taylor, J., & Turner, R. J. (2002). Perceived discrimination, social stress, and depression in the transition to adulthood: Racial contrasts. *Social Psychology Quarterly, 65*(3), 213-225.

Race, Racism, Ethnicity, Racial Discrimination, & Related Measures

TITLE OF MEASURE	RACE-RELATED STRESS
Contact Information	David Williams Department of Sociology University of Michigan Ann Arbor, MI 48106-1248, USA Tel: 734-936-0649 e-mail: wildavid@isr.umich.edu

Sexism & Sex Discrimination

Sexism & Sex Discrimination

TITLE OF MEASURE	SEXIST ATTITUDES TOWARD WOMEN SCALE (SATWS)
Source/Primary reference	Benson, P., & Vincent, S. (1980). Development and validation of the Sexist Attitudes Toward Women Scale (SATWS). *Psychology of Women Quarterly, 5,* 276-291.
Construct measured	Tendency toward and support for sexist attitudes
Brief description	This scale includes 40 items. Each item is rated on a 7-point scale from "strongly agree" to "strongly disagree" (high score = high sexism). The items concern 6 content areas: 1. Attitudes that women are genetically inferior (emotionally, biologically, intellectually) to men 2. Belief for the premise that men are entitled to greater power, prestige, and social advantage 3. Hostility toward women who engage in traditionally masculine roles and behaviors or who fail to fulfill traditional female roles 4. Lack of support and empathy for the women's liberation movement and the issues involved in such a movement 5. Use of derogatory labels and restrictive stereotypes in describing women 6. Evaluation of women on the basis of attractiveness information and willingness to treat women as sexual objects 24 items are sexist remarks and 16 are non-sexist ones (requiring inverse scoring).
Sample items	- I think that men are instinctually more competitive than women. - I see nothing wrong with men who are primarily interested in a women's body.
Appropriate for whom (i.e. which population/s)	Adult women and men
Translations & cultural adaptations available	None known
How developed	The authors define sexist attitudes toward women as "attitudes that function to place females in a position of relative inferiority to males

Sexism & Sex Discrimination

TITLE OF MEASURE	SEXIST ATTITUDES TOWARD WOMEN SCALE (SATWS)

by limiting women's social, political, economic, and psychological development" (p. 278). Items were written by the authors to reflect multiple hypothesized dimensions of sexism toward women, then refined through pilot testing.

On the basis of feminist literature and discussion with feminists, the authors identified 7 components of sexism toward women. Then, together with 3 colleagues they wrote 20-21 items to assess each of these 7 components. The resulting 141 items were administered to a development sample:

Development Sample		Demographics
Sample Size		n = 886
Description		482 college students; 402 non-college adults
Gender	Female	n = 487
	Male	n = 399
Race/Ethnicity		Not Reported

As a result, 91 items were retained from the original poll of 141 items. Two of the original 7 components were merged together, thus obtaining the 6 components included in the scale. From the pool of 91 items, the authors chose 10 items for each component, obtaining a 60-item scale. Using the data from the original development sample of 886 people, the authors performed scale intercorrelations. As a result, the 60 items were collapsed into the final single 40-item scale. Using again the data from the development sample relative to the 40 retained items, Cronbach's α was calculated to assess SATWS internal consistency. The coefficient obtained was very high: .91.

Psychometric properties

STUDY SAMPLES

Participants			Study 1		Study 2
Sample Size & Description		Students	n = 80		n = 58
		Non-students	n = 72		Non-student adults
Age	Range		28-74		-
	Mean		-		42.7
Gender	Students				
	Female		n = 40		-
	Male		n = 40		-
	Non-students				
	Female		n = 38		n = 30
	Male		n = 34		n = 28
Race/Ethnicity			Not reported		Not reported

TITLE OF MEASURE — SEXIST ATTITUDES TOWARD WOMEN SCALE (SATWS)

Sexism & Sex Discrimination

VALIDITY

Content Validity

The authors point out that the scale content validity is enhanced by the fact that it covers a wider range of content areas than other scales that measure sex-role stereotypes or attitudes toward women.

The SATWS was not contaminated by social desirability as measured by the Marlowe-Crowne Social Desirability Scale (Crowne & Marlowe, 1960).

Construct Validity

Overall, SATWS appears to have a very good construct validity:

- It was correlated as expected with attitudinal and behavioral self-report measures in other domains (e.g., literature preferences, driving frequency relative to spouse/partner/lover, making personality attributions as a function of physical appearance).

- It was not correlated with constructs where not expected: social responsibility, creativity, and social desirability (divergent validity).

Concurrent Validity

The SATWS was correlated with other scales that seek to measure similar constructs:

- The Attitudes Toward Women Scale (AWS; Spence & Helmreich, 1972)

- Sex-role stereotypes as measured by a short form of the Personal Attributes Questionnaire (PAQ; Spence, Helmreich, & Stapp, 1974)

- Support for the women's liberation movement (Women's Liberation Movement Scale - WLM; Tavris, 1973)

Scale	SATWS
ATWS	.36**
PAQ	-.65**
WLM	.68**

**$p<.01$

Sexism & Sex Discrimination

TITLE OF MEASURE	SEXIST ATTITUDES TOWARD WOMEN SCALE (SATWS)

RELIABILITY

Internal Consistency

The SATWS had high internal consistency and reliability for both college students and nonstudent adults:

Scale	Student Sample $\alpha =$	Non-student Sample $\alpha =$
SATWS	.90	.93

Comments	- SATWS appears to be a better measure for sexism than scales that assess only one or two of the components of sexism.
	- Internal consistency, content validity, and construct validity of SATWS were very good; no data were available for test-retest reliability.
	- The ethnic/racial make-up of the sample was not reported. It would be useful to assess its validity and reliability for multiple ethnic/racial groups.

Bibliography (studies that have used the measure)	Schram, P. (1998). Stereotypes about vocational programming for female inmates. *Prison Journal, 78*(8) 244.

Contact Information	Peter L. Benson
	Search Institute
	Thresher Sq West
	700 S. Third St.
	Minneapolis, MN 55415, USA
	Tel: 612-376-8955
	e-mail: peterb@search-institute.org

Sexism & Sex Discrimination

TITLE OF MEASURE	THE AMBIVALENT SEXISM INVENTORY (ASI)
Source/Primary reference	Glick, P., & Fiske, S. T. (1996). The ambivalent sexism inventory: Differentiating hostile and benevolent sexism. *Journal of Personality and Social Psychology, 70*(3), 491-512.
Construct measured	Hostile and benevolent sexism toward women
Brief description	The ASI consists of 22 items divided into two subscales: 1. Hostile sexism subscale covers three categories: <ul><li>Dominative paternalism</li><li>Competitive gender differentiation</li><li>Heterosexual hostility</li></ul> 2. Benevolent sexism subscale covers three categories: <ul><li>Protective paternalism</li><li>Complementary gender differentiation</li><li>Heterosexual intimacy</li></ul> Each subscale consists of 11 items and is rated on a 6-point rating scale from 0 = disagree strongly to 5 = agree strongly.
Sample items	Hostile Sexism (HS): <ul><li>The world would be a better place if women supported men more and criticized them less.</li><li>A wife should not be significantly more successful in her career than her husband.</li><li>There are many women who get a kick out of teasing men by seeming sexually available and then refusing male advances.</li></ul> Benevolent Sexism (BS): <ul><li>Every woman should have a man to whom she can turn for help in times of trouble.</li><li>Many women have a quality of purity that few men possess.</li><li>People are not truly happy in life unless they are romantically involved with a member of the other sex.</li></ul>
Appropriate for whom (i.e. which population/s)	Adults

Sexism & Sex Discrimination

TITLE OF MEASURE	THE AMBIVALENT SEXISM INVENTORY (ASI)
Translations & cultural adaptations available	There are multiple versions: Turkish, German, Dutch, Italian, Spanish, Portuguese, Chinese, Korean, Japanese

How developed

The researchers developed 140 statements to represent the conceptual categories derived from their theoretical analysis. Items were included to assess subjective positive feelings men have toward women. Nine items were adapted from Katz and Hass's (1988) Pro-black Scale, converting the target group to women (e.g., Women do not have the same employment opportunities that men do). Several items expressed recognition of continuing discrimination against women (e.g., Popular culture is very sexist). Six obviously correct/incorrect statements were included to assess validity and response biases (e.g., Few secretarial jobs are held by women).

Based on the results of an initial study, items with extreme means, based on cutoffs of 1or less and 4 or more, were excluded. Items excluded included the 6 validity items and 22 other items. The remaining 22 items were chosen on the basis of:

a. the items' tendency to load consistently highly on the HS and BS factors that emerged in the separate factor analyses for men and women.

b. maintaining diversity in the various aspects of sexism apparently tapped by the items.

c. consistent performance by the items in subsequent studies (Studies 1 to 4 described below).

Psychometric properties

STUDY SAMPLES

Participants		Study 1	Study 2	Study 3
Sample Size		n = 833	n = 171	n = 937
Description		Students from 3 universities; 2 in MA & 1 in Midwest	Students from 1 university in MA	Students from 1 university in MA
Age	Mean	19.5-20.7	Similar to Study 1†	Similar to Studies 1 & 2
Gender	Female Male	n = 480 n = 353	n = 94 n = 77	n = 541 n = 396
Race/Ethnicity	White Asian	76-86%‡	Similar to Study 1†	81%‡ 6%

†Authors state that although age and ethnicity were not recorded, the sample appears to be similar to the sample in Study 1.

‡ No additional race/ethnicity detail is reported.

TITLE OF MEASURE

Sexism & Sex Discrimination

THE AMBIVALENT SEXISM INVENTORY (ASI)

Age	Range	18-77		Similar to Study 2†
	Median	34		
	Participants	Study 4	Study 5	Study 6
	Sample Size	n = 144	n = 112	n = 85
	Description	Non-student adults recruited in MA	Non-student adults recruited in MA and Midwest	Students from 1 university in MA
Gender	Female Male	n = 72 n = 72	n = 76 n = 36	n = 41 n = 44
Race/ Ethnicity	White	83%‡		Similar to Study 2†

†Authors state that sample 6 is similar in characteristics to the sample in Study 2.

‡No additional race/ethnicity detail is reported.

VALIDITY

Content Validity

Study 2: In order to control for socially desirable responses, participants completed both the ASI and the Balanced Inventory of Desirable Responding (BIDR; Paulhus, 1988). ASI was not related to the Self-Deception scale but showed a significant correlation with the BIDR Impression Management scale.

Construct Validity

To assess construct validity of the two scales, participants in Studies 4-6 were given a semantic differential scale used by Eagly et al. (1991) to measure attitudes toward specific social groups. The authors predicted that HS would be correlated with negative attitudes toward women and BS would be correlated with positive attitudes toward women. As expected, the overall ASI score did not predict general attitudes toward women. Also as predicted, for men in the nonstudent samples (Studies 4 and 5) the two subscales had the opposite relationships to attitude toward women. In these two studies, the more men expressed positive attitudes toward women, the more benevolent and the less hostile sexism they expressed. Corresponding correlations were not found in the student sample (Study 6), and the results for women were less consistent.

Sexism & Sex Discrimination

TITLE OF MEASURE — THE AMBIVALENT SEXISM INVENTORY (ASI)

Expected differences between women and men were found in all six studies (all F's > 4.82, $p<.05$), with men scoring higher than women on both subscales.

Concurrent Validity

In the second study, participants completed four scales that measure sexism and hostility toward women:

- Attitudes Toward Women Scale (AWS; Spence & Helmreich, 1972)
- Old-Fashioned Sexism Scale (Swim et al., 1995)
- The Modern Sexism Scale (Swim et al., 1995)
- Rape Myth Acceptance Scale (RMA; Burt, 1980)

The ASI showed moderate correlations with most of the other measures of sexism:

ASI Scale	AWS	Old-fashioned sexism	Modern sexism	RMA
ASI	.63**	.42**	.57**	.54**
HS	.68**	.48**	.65**	.61**
BS	.40**	.24**	.33**	.32**
Controlling for Impression Management				
ASI	.61**	.38**	.62**	.54**
HS	.67**	.44**	.70**	.61**
BS	.38**	.19**	.36**	.31**
HS controlling for BS	.60**	.43**	.60**	.55**
BS controlling for HS	.04	-.03	-.06	-.02

**$p < .01$

Cross-Cultural Validity

Glick et al. (2000) administered the ASI in 19 nations. The complex factor structure of the ASI (HS and BS with 3 subfactors) was replicated in confirmatory factor analyses (the preferred model was the best fit in 16 of the 19 nations). In 12 nations in which spontaneous stereotypes of women were measured, HS predicted negative and BS predicted positive valences in stereotypic traits. Despite the relationship of BS to subjectively positive images of women, national averages on BS (as well as on HS) scores were negatively related to national indicators of gender equality.

Sexism & Sex Discrimination

TITLE OF MEASURE — THE AMBIVALENT SEXISM INVENTORY (ASI)

RELIABILITY

Internal Consistency

Scale	Study 1 α =	Study 2 α =	Study 3 α =	Study 4 α =	Study 5 α =	Study 6 α =
ASI	.92	.88	.83	.83	.87	.90
HS	.92	.87	.80	.87	.91	.89
BS	.85	.75	.77	.78	.73	.83

The Benevolent Sexism subscale consistently presents lower alpha coefficients, which can be explained by the limited number of items, considering its multidimensional nature.

Comments

- In cross-national comparisons, BS and HS negatively predict national indicators of gender equality (which included health-related measures such as gender differences in longevity).

- Has been shown to be reliable and valid for cross-cultural use.

- The U.S. study samples were predominantly white. It would be useful to have more information about the scale's validity and reliability for multiple ethnic/racial groups within the U.S.

Bibliography (studies that have used the measure)

Glick, P., et al. (2000). Beyond prejudice as simple antipathy: Hostile and benevolent sexism across cultures. *Journal of Personality and Social Psychology, 79*, 763-775.

Glick, P., Diebold, J., Bailey-Werner, B., & Zhu, L. (1997). The two faces of Adam: Ambivalent sexism and polarized attitudes toward women. *Personality and Social Psychology Bulletin, 23*, 1323-1334.

Glick, P., & Fiske, S. T. (1996). The Ambivalent Sexism Inventory: Differentiating hostile and benevolent sexism. *Journal of Personality and Social Psychology, 70*, 491-512.

Glick, P., & Fiske, S. T. (2001). An ambivalent alliance: Hostile and benevolent sexism as complementary justifications for gender inequality. *American Psychologist, 56*(2), 109-118.

Glick, P., Sakalli-Ugurlu, N. & Ferreira, M. (2002). Ambivalent sexism and attitudes toward wife abuse in Turkey and Brazil. *Psychology of Women Quarterly, 26*(4), 292-297.

Contact Information

No cost, but permission to use the AMI is required for commercial uses. Contact Peter Glick.
e-mail: glickp@lawrence.edu

Sexism & Sex Discrimination

TITLE OF MEASURE	THE AMBIVALENCE TOWARD MEN INVENTORY (AMI)
Source/Primary reference	Glick, P., & Fiske, S. T. (1999). The ambivalence toward men inventory. *Psychology of Women Quarterly, 23*(3), 519-536.
Construct measured	Women's hostile and benevolent prejudices toward men
Brief description	The AMI consists of 20 items divided into two subscales: 1. Hostility toward men (HM) 2. Benevolence toward men (BM) Each subscale addresses three subfactors of male structural power: 1. paternalism/maternalism 2. gender differentiation 3. sexuality The responses on a 6-point rating scale range from 0 = strongly disagree to 5 = strongly agree.
Sample items	Hostility toward Men - Men will always fight for greater control in society. - Most men are really like children. - When in positions of power, men sexually harass women. Benevolence toward Men - Even if both work, women should take care of men at home. - Men are more willing to risk self to protect others. - Every woman needs a male partner who will cherish her.
Appropriate for whom (i.e. which population/s)	Adults
Translations & cultural adaptations available	Has been used successfully in cross-cultural work (Glick et al., 2003)
How developed	This instrument aims to measure women's hostile and benevolent prejudices and stereotypes about men. Theoretical analysis led the authors to distinguish between two dimensions of the phenomenon: 1) ***Hostility toward men*** (which taps Resentment of paternalism, Compensatory

TITLE OF MEASURE: **THE AMBIVALENCE TOWARD MEN INVENTORY (AMI)**

Sexism & Sex Discrimination

gender differences, and Heterosexual hostility) and 2) **Benevolence toward men** (which taps Materialism, Complementary gender differences, and Heterosexual intimacy).

The authors conducted three studies to develop AMI. In the first, respondents answered 133 questions rated on a five-point Likert-type scale. Many of these items were inspired by discussions of a small group of women who recorded their attitudes toward men in the absence of the researchers. Study 1 was used to select 32 items that most cleanly loaded on the separate factors, and would be used in the following studies. Further analysis reduced the items of the scale to 20.

Psychometric properties

<u>STUDY SAMPLES:</u> Three studies established AMI's psychometric properties.

Participants Sample Size		Study 1 $n = 480$	Study 2 $n = 208$	Study 3 $n = 266$
Description		Students from 3 universities; 2 in MA & 1 in Midwest	Students from 1 university in MA	Nonstudent adults from the Midwest
Age	Range	>90% 17 – 24	Similar to Study 1†	16 – 86
	Median Women	‡	‡	44
	Median Men	‡	‡	48
Gender	Female	$n = 333$	$n = 134$	$n = 164$
	Male	$n = 147$	$n = 74$	$n = 102$
Race/ Ethnicity	White	86.5%	Similar to Study 1†	95%
	Asian	4%		3%
	Hispanic	1.4%		<1%
	Native American	1%		‡
	African American	1.2%		1.4%
	Unspecified	5.9%		‡

†Authors state that the sample is similar to the sample in Study 1.
‡Not reported

<u>VALIDITY</u>

Construct Validity

The complex structure of the AMI (HM and BM subscales each with 3 subfactors) was replicated in 11/12 nations with sufficient sample size for confirmatory factor analysis (Glick et al., 2003). HM (negatively)

Sexism & Sex Discrimination

TITLE OF MEASURE | THE AMBIVALENCE TOWARD MEN INVENTORY (AMI)

and BM (positively) predicted the valence of stereotypes toward men in the 6 nations in which this has been tested (Glick & Fiske, 1999; Glick et al., 2003).

RELIABILITY

Internal Consistency

Internal consistency was established in Studies 1 through 3.

Scale	Cronbach's α Range
HM	.81 - .86
BM	.79 - .83
Overall	.83 - .87

The AMI scales are highly reliable. Average alpha coefficients in a 16-nation study were .76 for HM and .77 for BM (Glick et al., 2003).

Comments
- Average HM and BM scores are negatively related to national indicators of gender equality in cross-national comparisons, which include measures, such as longevity, that are related to health (Glick et al., 2003). These data suggest that HM, despite being associated with negative stereotypes of men, justifies gender inequality (by characterizing men as arrogant, yet powerful).
- Good evidence of cross-cultural reliability and validity.

Bibliography (studies that have used the measure)

Glick, P., & Fiske, S. T. (1999). The ambivalence toward men inventory. *Psychology of Women Quarterly, 23*(3), 519-536.

Glick, P., Lameiras, M., & Castro, Y. R. (2002). Education and Catholic religiosity as predictors of hostile and benevolent sexism toward women and men. *Sex Roles, 47*, 433-441.

Glick, P. et al. (2003). Hostile as well as Benevolent Attitudes Toward Men Predict Gender Hierarchy: A 16-Nation Study. Lawrence University, Appleton, WI. Manuscript submitted for publication.

Contact Information

No cost, but permission to use the AMI is required for commercial uses. Contact Peter Glick.
e-mail: glickp@lawrence.edu

Sexism & Sex Discrimination

TITLE OF MEASURE	SCHEDULE OF SEXIST EVENTS (SSE)
Source/Primary reference	Klonoff, E. A., & Landrine, H. (1995). The schedule of sexist events. *Psychology of Women Quarterly, 19*(4), 430-472.
Construct measured	Lifetime and recent sexist discrimination in women's lives
Brief description	The SSE is a self-report inventory containing 20 items that are each rated in three different ways: once for the frequency in the last year, another time for the frequency in the respondent's lifetime, and a third time for appraising the stressfulness of each event. Response options range from 1 = the event never happened to me, to 6 = the event happens all of the time for the first two subscales, and 1 = not at all stressful to 6 = very stressful, for the third subscale.
Sample items	▪ How many times have you been treated unfairly by your employer, boss or supervisor because you are a woman? ▪ How many times were you forced to take drastic steps (such as filing a grievance, filing a lawsuit, quitting your job, moving away and other actions) to deal with some sexist thing that was done to you?
Appropriate for whom (i.e. which population/s)	Adult women
Translations & cultural adaptations available	None known
How developed	The authors developed 23 items for the scale based on input from women in a variety of contexts. Women were approached on a college campus, in nine small office buildings, and while waiting in a local airport and were asked to complete an anonymous survey. During the data analysis, three items did not load on any factor, so they were omitted from the scale, yielding a final scale with 20 items.
Psychometric properties	*STUDY SAMPLE*

Participants		Demographics
Sample Size		n = 631
Age	Range	18-73
	M (SD)	32.14 (11.74)
	Median	29
Gender	Female	100%

Sexism & Sex Discrimination

TITLE OF MEASURE — SCHEDULE OF SEXIST EVENTS (SSE)

Participants		Demographics
Race/Ethnicity	Black	$n = 38$
	Latina	$n = 117$
	Asian American	$n = 25$
	White	$n = 403$
	Other	$n = 46$
Education	College or Graduate Degree	$n = 119$
	Some College	$n = 340$
	High School or Less	$n = 129$
Income	Range	$0 - $400,000
	M (SD)	$34,058 ($34,370)
Marital Status	Single	$n = 292$
	Married	$n = 238$
	Widowed/Separated/Divorced	$n = 101$

VALIDITY

Construct Validity

The data were entered in a principal components analysis with an orthogonal rotation and factors retained based on an eigenvalue equal to or greater than 1.00. Four factors emerged from the analysis, accounting for 58.8% of the variance:

- Sexist degradation and its consequences (I)
- Sexism in distant relationships (II)
- Sexism in close relationships (III)
- Sexist discrimination in the workplace (IV)

Concurrent Validity

SSE was compared to two measures of frequency of stressful events: the Psychiatric Epidemiology Research Interview-Life Events Scale (PERI-LES; Dohrenwend, Krasnoff, Askenasy, & Dohrenwend, 1978) and the Hassles Frequency Scale (Hassles–F; Kanner, Coyne, Schaeffer, & Lazarus, 1981).

Scale	SSE–Recent $r =$	PERI–LES $r =$	Hassles–F $r =$
SSE–Lifetime	.75	.27	.24
SSE–Recent		.27	.24
PERI–LES			.32

Note: All correlations are significant at $p < .00005$

Sexism & Sex Discrimination

TITLE OF MEASURE — SCHEDULE OF SEXIST EVENTS (SSE)

	SSE Factors				
Scale	I $r=$	II $r=$	III $r=$	IV $r=$	TOTAL $r=$
Lifetime					
Hassles–F	.21	.19	.22	.22	.24
PERI–LES	.26	.23	.21	.17	.27
Recent					
Hassles–F	.19	.20	.17	.23	.29
PERI–LES	.28	.15	.19	.18	.27

RELIABILITY

Internal Consistency

Internal consistency reliability was calculated for all the subscales of the SSE Lifetime and Recent

SSE–Lifetime	Cronbach's α =
Sexist degradation and its consequences	.89
Sexism in distant relationships	.82
Sexism in close relationships	.67
Sexist discrimination in workplace	.68
TOTAL	.92

SSE–Recent	Cronbach's α =
Sexist degradation and its consequences	.88
Sexism in distant relationships	.74
Sexist discrimination in workplace	.70
Sexism in close relationships	.61
TOTAL	.90

Test-Retest Reliability

Test-retest reliability was not considered an adequate way to assess the reliability of the scales since a single event occurring on a day would change the scores of both scales. However, a test-retest analysis was conducted with a small sample of 50 college women, over an interval of two weeks.

Scale	$r=$
SSE-Lifetime	.70**
SSE-Recent	.63**

**$p < .01$

Sexism & Sex Discrimination

TITLE OF MEASURE	SCHEDULE OF SEXIST EVENTS (SSE)

Split-half reliability was deemed to be the best way to assess the test-retest reliability of both SSE-Lifetime and SSE-Recent scales.

Scale	r =
SSE-Lifetime	.87***
SSE-Recent	.83***

***$p < .001$

Comments

- Evidence that experiences of sexism as measured by the SSE are related to both physical and mental health of women.

- The sample contained only a small number of African American and Asian American women. The factor structure might have been different if the sample were more ethnically and racially diverse.

- The lack of information about the women's appraisal of the stressfulness of sexist events is a limitation of the scale.

Bibliography (studies that have used the measure)

Klonoff, E., Landrine, H., & Campbell, R. (2000). Sexist discrimination may account for well-known gender differences in psychiatric symptoms. *Psychology of Women Quarterly, 24*, 93-99.

Landrine, H., Klonoff, E., Gibbs, J., Manning, V., & Lund, M. (1995). Physical and psychiatric correlates of gender discrimination: An application of the Schedule of Sexist Events. *Psychology of Women Quarterly, 19*, 473-492.

Yoder, J., & McDonald, T. (1998). Measuring sexist discrimination in the workplace: Support for the validity of the Schedule of Sexist Events. *Psychology of Women Quarterly, 22*, 487-491.

Contact Information

Elizabeth A. Klonoff
Department of Psychology
San Diego State University
5500 Campanile Drive
San Diego, CA 92182-4611, USA

Sexism & Sex Discrimination

TITLE OF MEASURE	STIGMA CONSCIOUSNESS QUESTIONNAIRE (SCQ)
Source/Primary reference	Pinel, E. C. (1999) Stigma consciousness: The psychological legacy of social stereotypes. *Journal of Personality and Social Psychology, 76*(1), 114-128.
Construct measured	The extent to which people focus on their stereotyped status
Brief description	The SCQ (originally designed for use with women) can be modified for use with any stereotyped group. It predicts perceptions of discrimination as well as many negative consequences of discrimination (e.g., impaired performance, disidentification, and lowered self-esteem). The SCQ consists of 10 items on a rating scale from 0 = disagree strongly to 6 = agree strongly. 7 items are reverse scored.
Sample items	Examples from SCQ for Women: - Stereotypes about women have not affected me personally. - When interacting with men, I feel like they interpret all my behaviors in terms of the fact that I am a woman. - Most men have a lot more sexist thoughts than they actually express.
Appropriate for whom (i.e. which population/s)	Adults, although one could theoretically modify the scale for use with children.
Translations & cultural adaptations available	None currently
How developed	The initial version of the scale was the SCQ for women. The 16 original items were written to span two broad content areas: 1) women's phenomenological experiences when interacting with men, and 2) beliefs about how men view women. At the end of Study 1, 10 items of the original questionnaire were retained. Studies 2, 3, 4, and 5 provide evidence for the scale's psychometric properties. Studies 3 and 4 tested the generalizability of the construct to gays and lesbians and ethnic/racial minority groups. Study 6 illustrated some consequences of stigma consciousness.

Sexism & Sex Discrimination

TITLE OF MEASURE   STIGMA CONSCIOUSNESS QUESTIONNAIRE (SCQ)

Psychometric properties

<u>STUDY SAMPLES</u>

Participants		Study 1	Study 2	Study 3
Sample Size		n = 753	n = 86 (phase 1) n = 57 (phase 2)	n = 66
Description		Female introductory psychology students	Female introductory psychology students. 44 women who completed phase 1 participated in phase 2.	Gay men and lesbians recruited at the 1997 Gay Pride Festival, San Diego, California
Age	Mean	19.5-20.7	19	not known
Gender	Female	n = 753	n = 94	n = 27
	Male	n = 0	n = 77	n = 23
Race/Ethnicity (when indicated)	White	n = 472	n = 46	
	Black	n = 62	n = 6	
	Asian	n = 83	n = 8	
	Hispanic	n = 101	n = 11	
	Native American	n = 4	n = 4	

Participants		Study 4	Study 5	Study 6
Sample Size		n = 337	n = 393	n = 81
Description		Introductory psychology students	23 gay men and 27 lesbians who participated in Study 3. 142 men, 201 women, 200 Whites and 21 Blacks who participated in Study 4.	Female college students
Age	Mean	Not known	Not known	Similar to Studies 1 & 2
Gender	Female	n = 201	n = 228	n = 541
	Male	n = 136	n = 165	n = 396
Race/Ethnicity (when indicated)	White	n = 198	n = 200	81%
	Black	n = 21	n = 21	6%
	Asian	n = 63		
	Hispanic	n = 53		
	Native American			

Sexism & Sex Discrimination

TITLE OF MEASURE: STIGMA CONSCIOUSNESS QUESTIONNAIRE (SCQ)

VALIDITY

Construct Validity

Study 1: A principal-axis factor analysis with varimax rotation was conducted for the initial 16 SCQ items. Only one factor with an eigenvalue of 2.92 emerged, accounting for 83% of the common variance and 11% of the total variance. 10 SCQ items that loaded .33 or higher on the single factor were retained.

Another principal-axis factor analysis was conducted for the retained 10-item scale. Again, one factor emerged accounting for 96.5% of the common variance and 24% of the total variance. All 10 items loaded .32 or higher on the single factor, with 0.48 average loading value.

The 10-item SCQ scale was administered to a new sample of 302 female introductory psychology students. Again, principal-axis factor analysis was conducted, and, similarly, one factor was revealed that accounted for 91% of the common variance and 23% of the total variance.

Study 2: To provide further evidence for construct validity, correlations between stigma consciousness, as measured in both Phase 1 and Phase 2, and various measures of discrimination were computed. Women high in stigma consciousness are more likely than women low in stigma consciousness to perceive discrimination at the group, average, and personal levels.

Measure	SCQ	
	Phase 1	Phase 2
Group Discrimination	.36*[a]	.48**[b]
Average Discrimination	.33*[a]	.50**[c]
Personal Discrimination	.37*[a]	.48**[b]

[a]$n = 44$; [b]$n = 57$; [c]$n = 56$
*$p < .05$; **$p < .01$

Study 3: SCQ was adapted for gay men and lesbians. A factor analysis resulted in one factor accounting for 74% of common variance.

Study 4: One of the goals of Study 4 was to examine whether the stigma consciousness construct is distinct from those of group identity and group consciousness. Factor analysis was conducted on SCQ for

Sexism & Sex Discrimination

Women, Sensitivity to Sexism scale (Henderson-King & Steward, 1997), and Revelation and Embeddedness subscales of Rickard's (1987) Feminist Identity scale. This analysis yielded four factors with eigenvalues of greater than 2. Consistent with the claim that stigma consciousness represents a unique factor, items 1-7 of the SCQ for Women loaded .3 or higher on one single factor and only one of these seven items loaded on any of the other factors. Two SCQ items loaded on the factor associated with items from the Revelation subscale, and the one remaining factor loaded on the factor associated with items from the Sensitivity to Sexism.

Concurrent Validity

Study 2: Various other scales were administered to examine whether the SCQ for Women correlates with measures that reveal how women who are high in stigma consciousness (i) express concern over how others view them and (ii) are attentive to signs of sexism. Self-Consciousness Scale (SCS-scale; Fenigstein, Scheier, & Buss, 1975) and Modern Sexism Scale (Swim, Aikin, Hall, & Hunter,1995) were administered. SCS-scale consists of three subscales: a Private Self-Consciousness subscale, a Public Self-Consciousness subscale, and a Social Anxiety subscale.

Correlations between the SCQ for Women with other measures described above

Measure	SCQ $r =$
Modern Sexism Scale	-.28 **
Private Self-Consciousness	.13
Public Self-Consciousness	.36 *

$n = 86$; *$p < .05$; **$p < .01$

Study 3: To examine the validity of the SCQ for gay men and lesbians, it was administered along with Fenigstein et al.'s (1975) Self-Consciousness Scale (SCS) and four measures of perceived discrimination.

Sexism & Sex Discrimination

TITLE OF MEASURE — STIGMA CONSCIOUSNESS QUESTIONNAIRE (SCQ)

Correlations of the SCQ and other measures

Measure	SCQ for Gays and Lesbians $r =$
Private Self-Consciousness	.33 **
Public Self-Consciousness	.33 **
Group discrimination	
Lesbians	.34 **
Gay men	.33 **
Gay men and lesbians	.50 **
Personal discrimination	.57 **

Lesbian, gay men, and self-discrimination measures, $n = 62$. For group gay men and lesbian discrimination measure, $n = 61$.

** $p < .01$

Study 4: Men and women of five different races completed two SCQs, one pertaining to their race and one pertaining to their sex. Analyses similar to those in previous studies were conducted.

	SCQ for Sex	
Measure	Men $n = 136$ $r =$	Women $n = 198$ $r =$
Private Self-Consciousness	.23 **	.31 **
Public Self-Consciousness	.09	.30 **
Discrimination		
Group	.19**	.28**
Average	.22*	.29**
Personal	.29**	.36**

	SCQ for Race			
Measure	Whites $n = 197$ $r =$	Blacks $N = 21$ $r =$	Asians $n = 63$ $r =$	Hispanics $n = 53$ $r =$
Private Self-Consciousness	.24**	.06	.28*	.12
Public Self-Consciousness	.16*	.02	.23*	.25***
Discrimination				
Group	.31**	.54*	.35**	.51**
Average	.32**	.49*	.26*	.63*
Personal	.42**	.77**	.40**	.64**

Sexism & Sex Discrimination

TITLE OF MEASURE	STIGMA CONSCIOUSNESS QUESTIONNAIRE (SCQ)

RELIABILITY

Internal Consistency

Cronbach α coefficients of the SCQ for different studies

Scale	Study 1 (phase 1) α =	Study 1 (phase 2) α =	Study 3 α =	Study 5 α =	Study 6 α =
SCQ	.74	.72	.81	.87	.90

Test-Retest Reliability

Study 2: The correlation between stigma consciousness as measured during Phase 1 and stigma consciousness as measured during Phase 2 (average of 5 weeks after Phase 1) was computed: $r(42) = .76$, $p < .001$.

Comments

- Stigma consciousness is a potentially useful concept for understanding how people respond to others in the workplace.

- The studies suggest that the SCQ is a useful, valid, and reliable measure.

- The research suggests that stigma consciousness is a domain-specific construct. Knowing people's stigma consciousness levels with respect to one of their group memberships (e.g., gender) does not necessarily inform us about their stigma consciousness levels with respect to their other group memberships (e.g., race).

Bibliography (studies that have used the measure)

Pinel, E. C., & Paulin, N. (2005). Stigma consciousness in the workplace. *Basic and Applied Social Psychology 27*(4):345-352.

Pinel, E. C., Warner, L. R., & Chua, P. (2005). Getting there is only half the battle: Stigma consciousness and maintaining diversity in higher education. *Journal of Social Issues 61*(3):481-506.

Pinel, E. C. (2004). You're just saying that because I'm a woman: Stigma consciousness and attributions to discrimination. *Self and Social Identity, 3,* 39-51.

Brown, R. P., & Pinel, E. C. (2003). Stigma on my mind: Individual differences in the experience of stereotype threat. *Journal of Experimental Social Psychology, 39,* 626-633.

Pinel, E. C. (2002). Stigma consciousness in intergroup contexts: The power of conviction. *Journal of Experimental Social Psychology, 38,* 178-185.

Sexism & Sex Discrimination

TITLE OF MEASURE: STIGMA CONSCIOUSNESS QUESTIONNAIRE (SCQ)

Contact Information

Elizabeth C. Pinel
Associate Professor of Psychology
Department of Psychology
Pennsylvania State University
543 Moore Building,
University Park, Pennsylvania 16802-3106, USA
Tel: 814-863-1149

Fax: 814-863-7002

e-mail: ecp6@psu.edu

http://psych.la.psu.edu/faculty/pinel.html

Sexism & Sex Discrimination

TITLE OF MEASURE	WORKING ENVIRONMENT FOR WOMEN IN ACADEMIC SETTINGS
Source/Primary reference	Riger, S., Stokes, J., Raja, S., & Sullivan, M. (1997). Measuring perceptions of the work environment for female faculty. *The Review of Higher Education, 21*(1), 63-78.
Construct measured	Perceptions of attitudes toward women faculty in university settings
Brief description	The 35-item scale includes 3 subscales: 1. Differential treatment 2. Balancing work and personal obligations 3. Sexist attitudes and comments There are two types of questions. Most were in the agree-disagree format ranging from 1 = do not agree to 5 = strongly agree. For the other questions, respondents are asked to report using a 5-point scale where 1 = not at all likely and 5 = very likely. A short version with 15 items has been developed. The authors do not recommend using subscale scores with the short version.
Sample items	<ul><li>Female faculty are less likely than their male counterparts to have influence in departmental policies and administration.</li><li>Faculty make jokes or comments that are demeaning or degrading to women.</li><li>Male faculty are comfortable having lunch alone with a female faculty member.</li><li>Most faculty are supportive of female colleagues who want to reduce their workload for personal reasons.</li><li>In this department sex discrimination is a big problem.</li><li>Male faculty are not as comfortable serving as a mentor to a female faculty member as they are to a male faculty member.</li></ul>
Appropriate for whom (i.e. which population/s)	Adults working in academia The authors have also developed a parallel climate scale for use in corporate environments (see entry for Stokes, Riger, & Sullivan, 1997)
Translations & cultural adaptations available	None known

Sexism & Sex Discrimination

TITLE OF MEASURE — **WORKING ENVIRONMENT FOR WOMEN IN ACADEMIC SETTINGS**

How developed

Open-ended interviews were conducted with 20 female faculty of different disciplines and from many colleges and universities in the Chicago area to assess their perceptions of the climate for female faculty within their academic departments. Based on the responses, literature review, and previous work of the authors, 200 items were generated to assess 6 domains (dual standards and treatment, sexist attitudes and comments, informal socializing, balancing work and personal obligations, remediation policies and practices, and mentoring). This questionnaire was piloted with 10 faculty, which resulted in a version of the scale that comprised 89 items, 81 in an agree-disagree format, and 8 quotations which respondents rated in terms of how likely they would be to hear such comments in their department. About half of the items were worded positively, and the other half were worded negatively. Further analyses described below yielded a scale with 35 items.

Psychometric properties

STUDY SAMPLE

Participants		Demographics
Sample Size		n = 626
Description		Faculty members of 69 colleges and universities in the U.S. and Canada
Age	Range	27-91
	Mean for Women	45.8
	Mean for Men	49.1
Gender	Female	64%
	Male	36%
Race/Ethnicity	White	97%
Employed Full-Time		98%
Rank		About equal numbers of assistant, associate, and full professors

VALIDITY

Construct Validity

Principal components analyses of the 35 items yielded three components that together accounted for 54.3% of the total variance. The first component, Differential Treatment ($\alpha = .95$), included 20 items from several of the a priori dimensions and seemed to be a general measure of the climate for women faculty. The second and third components were Balancing Work and Personal Obligations ($\alpha = .86$) and Sexist Attitudes and Comments ($\alpha = .85$). These results were not parallel to the Working Environment for Women in Corporate Settings developed by the same authors, where results confirmed a 5-factor solution (Stokes,

Sexism & Sex Discrimination

TITLE OF MEASURE — **WORKING ENVIRONMENT FOR WOMEN IN ACADEMIC SETTINGS**

Riger, & Sullivan, 1995). The authors speculated that this may be related to differences in distinct status categories, formality of hierarchy, and the fluidity of communication in corporate versus academic work environments.

RELIABILITY

Internal Consistency

Scale Form	α =
Long Form	.97
Short Form	.94

Correlation between short and long forms: .97

Very similar alphas were replicated with a validation sample (1/3 of the sample was analyzed separately to serve as a validation sample).

Subscale	α =
Differential Treatment	.96
Balancing Work and Personal Obligations	.83
Sexist Attitudes and Comments	.96

Comments

- The instrument is concise, easy to understand, and easy to administer.
- The scale was developed with a predominantly white sample. It would be useful to assess its validity and reliability for multiple ethnic/racial groups.

Bibliography (studies that have used the measure)

Contact Information

Stephanie Riger
Department of Psychology
University of Illinois at Chicago
1007 W. Harrison Street
Chicago, IL 60607, USA

Sexism & Sex Discrimination

Title of measure	**Working Environment for Women in Corporate Settings**
Source/Primary reference	Stokes, J., Riger, S., & Sullivan, M. (1995). Measuring perceptions of the working environment for women in corporate settings. *Psychology of Women Quarterly, 19*(4), 533-549.
Construct measured	Perceptions of attitudes toward women in the work environment
Brief description	The 36-item scale includes five subscales plus 4 global discrimination items: 1. Dual standards and opportunities (10 items) 2. Sexist attitudes and comments (7 items) 3. Informal socializing (7 items) 4. Balancing work and personal obligations (9 items) 5. Remediation policies and practices (3 items) There are two types of questions. Most were in the agree-disagree format ranging from 1 = do not agree to 5 = strongly agree. For the other questions, respondents are asked to report in a 5-point scale (1 = not at all likely, 5 = very likely) about the possibility that the quotations presented would be heard in their workplace. A short version with 15 items has been developed. The authors do not recommend using subscale scores with the short version.
Sample items	- Compared to men, women in this office are appointed to less important committees and task forces. - Jokes that are demeaning or degrading to women are told occasionally in this office. - Company-sponsored social events generally appeal to both female and male employees. - In general, supervisors in this company are understanding when personal or family obligations occasionally take an employee away from work. - People who raise issues about the treatment of women in this company are supported by other employees.
Appropriate for whom (i.e. which population/s)	Working adults The authors have also developed a parallel climate scale for use in academic environments (see entry for Riger, Stokes, Raja, & Sullivan)

Sexism & Sex Discrimination

TITLE OF MEASURE	WORKING ENVIRONMENT FOR WOMEN IN CORPORATE SETTINGS
Translations & cultural adaptations available	None known
How developed	Based on focus groups and feedback from men and women working in corporate environments, the authors developed both positively and negatively worded items for each of 6 hypothesized subscales:

1. Opportunities and mentoring

2. Inappropriate salience of gender

3. Sexist attitudes and comments

4. Informal socializing

5. Balancing work and personal obligations

6. Remediation policies and practices

A version of the questionnaire with 133 randomly ordered items (about half positively worded and half negatively worded) was completed by 398 people in 45 different companies. Analyses of these results yielded a final scale with 36 items.

Psychometric properties

<u>STUDY SAMPLE</u>

Participants		Demographics
Sample Size		n = 398
Age	Range	22-63
	Mean for Women	36.9
	Mean for Men	38.9
Gender	Female	n = 263
	Male	n = 134
Race/Ethnicity	White	92.4%
Education	Graduate or Professional Degree	51%
	Bachelor's Degree	91%
Income	$100,000 or more	40%
	$40,000 or more	90%
Marital Status	Never Married	25%
	Currently Married	65%
Children	No children under age 18 years	61%

TITLE OF MEASURE	*Sexism & Sex Discrimination* WORKING ENVIRONMENT FOR WOMEN IN CORPORATE SETTINGS
	<u>VALIDITY</u> **Construct Validity** One third of the surveys were randomly selected to be analyzed separately to serve as a validation sample. A series of principal component analyses yielded five subscales that were confirmed with the validation sample. <u>RELIABILITY</u> **Internal Consistency** | Scale	α =
Overall Scale	.96
Dual Standards and Opportunities	.92
Sexist Attitudes and Comments	.82
Informal Socializing	.82
Balancing Work and Personal Obligations	.90
Remediation Policies and Practices	.78
Short form	.93 | Correlation between short and long forms: .97 Values of α for the validation sample were almost identical to those reported above.
Comments	Designed specifically to assess workplace climate. - The instrument is concise, easy to understand, and easy to administer. - The scale was developed with a predominantly white sample. It would be useful to assess its validity and reliability for multiple ethnic/racial groups. - The availability of a short version makes it workable to include within a longer workplace survey. - Has been adapted to assess race-related climate (Bond, Punnett, Pyle, Cazeca, & Cooperman, in press; Yoder & Aniakudo, 1996). - Further psychometric research is needed, particularly due to the convenience sampling and lack of comparison with other measures of the work climate.
Bibliography (studies that have used the measure)	Bond, M. A., Punnett, L., Pyle, J. L., Cazeca, D., & Cooperman, M. (2004). Gendered work conditions, health, and work outcomes. *The Journal of Occupational Health Psychology, 9*(1), 28-45.

Sexism & Sex Discrimination

TITLE OF MEASURE	WORKING ENVIRONMENT FOR WOMEN IN CORPORATE SETTINGS

Yoder, J. D., & Aniakudo, P. (1996). When pranks become harassment: The case of African American women firefighters. *Sex Roles, 35*(5/6), 253-270.

Contact Information

Joseph Stokes
Department of Psychology (m/c 284)
University of Illinois at Chicago
1007 West Harrison Street
Chicago, IL 60607, USA

Sexism & Sex Discrimination

TITLE OF MEASURE	*OLD-FASHIONED AND MODERN SEXISM SCALE*
Source/Primary reference	Swim, J. K., Aikin, K. J., Hall, W. S., & Hunter, B. A. (1995). Sexism and racism: Old-fashioned and modern prejudices. *Journal of Personality and Social Psychology, 68*(2), 199-214.
Construct measured	Old-Fashioned Sexism (OFS) - endorsement of traditional gender roles, differential treatment of women and men, and stereotypes of women's lesser competence. Modern Sexism (MS) - denial of continued discrimination, antagonism toward women's demands, lack of support for policies to help women. This scale measures covert or subtle sexism, which is built into cultural or societal norms.
Brief description	The measure is a 13-item inventory with 2 subscales: 1. Old-Fashioned Sexism (5 items) 2. Modern Sexism (8 items) Each item is rated on a 5-point scale, with 1 = strongly disagree and 5 = strongly agree.
Sample items	Old-Fashioned Sexism: - Women are generally not as smart as men. - I would be equally comfortable having a woman as a boss as a man. Modern Sexism: - On average, people in our society treat husbands and wives equally. (denial of continued discrimination) - It is easy to understand the anger of women's groups in America. (antagonism toward women's demands) - Over the past few years, the government and news media have been showing more concern about the treatment of women than is warranted by women's actual experiences. (resentment regarding special favors for women).
Appropriate for whom (i.e. which population/s)	Adults
Translations & cultural adaptations available	None known

Sexism & Sex Discrimination

TITLE OF MEASURE	OLD-FASHIONED AND MODERN SEXISM SCALE

How developed

Theoretical background is that, similarly to racism, sexism can be seen as existing in two distinguishable forms: old-fashioned and modern. Based on the literature and past research, the authors wrote a set of statements to measure old-fashioned sexism. For modern sexism, they also wrote a set of statements based on three basic tenets that underlie the concept of "modern sexism": denial of continued discrimination, antagonism toward women's demands, and lack of support for policies to help women.

Psychometric properties

<u>STUDY SAMPLE</u>

Participants		Demographics
Sample Size		n = 683
Description		Undergraduate students from an introductory psychology class
Gender	Female	n = 418
	Male	n = 265
Race/Ethnicity		Nearly all respondents were European-American

<u>VALIDITY</u>
Construct Validity
The authors assessed the instrument's construct validity by performing a confirmatory factor analysis, investigating differences between female and male respondents' scores, testing the correlation between the scale and individualistic vs. egalitarian values, and then calculating the correction between the scale scores and perceptions of job segregation. Factor analyses supported the notion that OFS and MS are two distinct factors.

Campbell et al. (1997) compared the MS scale with the Neosexism scale. The scales correlated highly with each other, but most of the variance in one scale could not be explained by the variance in the other: thus the two instruments are not similar.

Swim and Cohen (1997) compared MS and OFS with the Attitudes Toward Women Scale (AWS, Spence & Helmreich, 1972), obtaining additional construct and discriminant validity for the MS scale. Confirmatory factor analyses show that AWS and OFS loaded on one factor that represents *overt* sexist beliefs. This factor is distinct from the factor that represents the MS scale. The MS scale seems to measure *covert, subtle* sexism. The OFS and AWS were more similar to each other than to the MS scale and their correlation was higher (.90 for both males and females) than the correlation between OFS and MS (.25 for males and .41 for females). MS was found to be a better predictor of sexual harassment than AWS (discriminant validity). AWS and MS correlate with affective reactions to different categories of men and women (general, traditional, feminists, and chauvinists) (convergent validity).

TITLE OF MEASURE	**Sexism & Sex Discrimination** *OLD-FASHIONED AND MODERN SEXISM SCALE*

However, MS and AWS are demonstrated to measure different, though related, constructs.

Men's scores on OFS and MS were higher than women's. A correlation matrix among OFS and MS scales, as well as Old-Fashioned Racism (OFR) and Modern Racism (MR) scales (McConahay, 1986) and both a shortened version of Mirels and Garrett's (1971) Protestant Work Ethic scale (PWE; Katz & Hass, 1988) and the Humanitarian-Egalitarian scale (HE; Katz & Hass, 1988), was derived separately for women and men. For women, the pattern of differences of the correlations between OFS, MS, OFR, and MR and PWE and HE were similar. That is, OFS, MS, OFR, and MR were each more strongly correlated with the HE scale than the PWE scale. For men, a similar pattern emerged with OFS and OFR, but not with MS or MR. This pattern of correlations provides partial support for the conclusion that modern prejudice is more strongly related to nonegalitarian beliefs than to highly individualistic beliefs (Sears, 1988). Higher scores on the MS scale correlated with overestimating women's presence in the workforce.

RELIABILITY

Internal Consistency

Scale	Cronbach's α Range
OFS	.65 - .66
MS	.75 - .84

Comments

- The scale was developed with a predominantly white sample. It would be useful to assess its validity and reliability for multiple ethnic/racial groups.

Bibliography (studies that have used the measure)

Campbell, B., Schellenberg, E. G., & Senn, C. Y. (1997). Evaluating measures of contemporary sexism. *Psychology of Women Quarterly, 21,* 89-102.

McHugh, M. C., & Frieze, I. H. (1997). The measurement of gender-role attitudes. A review and commentary. *Psychology of Women Quarterly, 21,* 1-16.

Swim, J. K., & Cohen, L. L. (1997). Overt, covert, and subtle sexism: A comparison between the Attitudes Toward Women and Modern Sexism Scales. *Psychology of Women Quarterly, 21,* 103-118.

Contact Information

Janet K. Swim
515 Moore Building
Department of Psychology
Pennsylvania State University
University Park, PA 16802, USA
e-mail: JKS4@PSUVM.PSU.EDU

Sexism & Sex Discrimination

TITLE OF MEASURE	EVERYDAY SEXISM
Source/Primary reference	Swim, J. K., Hyers, L. L., Cohen, L. L., & Ferguson, M. J. (2001). Everyday sexism: Evidence for its incidence, nature, and psychological impact from three daily diary studies. *Journal of Social Issues, 57*(1), 31-53.
Construct measured	Incidence, nature, and impact of everyday sexism
Brief description	Daily diary approach to recording both personal experiences with and observations of sexist events. For each incident observed, participants were asked to note the time the incident occurred, rate the impact it had on them on a scale ranging from -2 (very negative) to 0 (no impact) to +2 (very positive), and rate the extent to which the incident was sexist from -2 (definitely not sexist) to 0 (uncertain) to +2 (definitely sexist).
Sample items	Participants are told that their role is to record incidents where women are treated differently because of their gender. They are told to note incidents that are directed toward them, someone else, or women in general. In order to obtain a manageable number of incidents to record, participants are told to exclude observations from the media, such as television programming and advertisements. If they observe a gender-related incident, they are to complete the form as soon as possible after the incident has occurred. If more than one incident occurred on one day they are to complete a form for each incident. If they do not observe any gender-related incidents on any particular day, they are to note this on one of the forms at the end of each day.
Appropriate for whom (i.e. which population/s)	Adolescents and adults
Translations & cultural adaptations available	Swim has also used a similar diary approach for recording experiences of racism and differential treatment based on sexual orientation.
How developed	The authors designed the diary approach as an alternative to retrospective strategies.

Sexism & Sex Discrimination

TITLE OF MEASURE EVERYDAY SEXISM

Psychometric properties

STUDY SAMPLES

Participants Sample Size		Study 1 N = 40	Study 2 n = 37	Study 3 n = 73
Description		Students enrolled in an introductory psychology of gender course	Students enrolled in two introductory psychology courses and one advanced marketing course	Students enrolled in a psychology of gender course and their male friends
Age	Range	19 - 26	18 – 44	†
	Median	22	22	†
Gender	Female	n = 40	n = 20	n = 47
	Male		n = 17	n = 26
Race/ Ethnicity		†	†	†

†Not reported

VALIDITY

The authors argue that:

Much of the existing research on people's experiences with sexism is in the form of retrospective accounts in which participants are asked to characterize what they typically experience, sometimes for more than a year's worth of experiences. Such approaches often neglect more mundane "everyday" types of experiences and thus may provide an incomplete picture of the extent and variety of daily experiences with sexism. Retrospective surveys and interviews may not accurately reflect the extent and nature of experiences people have with prejudice for the following reasons. First, uncertainty about labeling subtle and ambiguous incidents as prejudicial may decrease the likelihood that such incidents are encoded and recalled as prejudicial. Second, isolated incidents may be minimized over time or seen as insignificant and therefore forgotten. Third, the similarity and commonness of incidents that constitute everyday prejudice may make it difficult to assess the frequency with which they occur through expansive retrospection. Finally, retrospective reports are subject to distortion as moods dissipate and contexts change. In contrast, daily diary studies minimize many of these problems, providing a more accurate and complete report of incidents and responses to them without the distorting processing that may result in errors.

Comments

- The authors found that sexist incidents as measured through diaries affected women's psychological well-being by decreasing their self-esteem.

Sexism & Sex Discrimination

TITLE OF MEASURE	EVERYDAY SEXISM

- The diary approach had the advantage of yielding qualitative data that can be quantified while at the same time being potentially richer in detail than survey data.

Bibliography (studies that have used the measure)

Contact Information

Janet K. Swim
515 Moore Building
Department of Psychology
Pennsylvania State University
University Park, PA 16802, USA

e-mail: JKS4@PSUVM.PSU.EDU

Sexism & Sex Discrimination

Title of measure	General Attitudes toward Affirmative Action (AA) and Men's Collective Interest
Source/Primary reference	Tougas, F., Brown, R., Beaton, A. M., & Joly, S. (1995). Neosexism: Plus ca change, plus c'est pareil. *Personality and Social Psychology Bulletin, 21*(8), 842-891.
Construct measured	Attitudes toward affirmative action
Brief description	The 11-item scale uses two approaches to measuring attitudes toward affirmative action: 1) General Attitudes (AA) The general scale includes 3 items to assess the respondents' attitudes toward affirmative action. Ratings are made on a 7-point scale where 1 = total disagreement and 7 = total agreement. Composite scores are calculated by taking a mean. 2) Impact on Men's Collective Interest (CI) Following a brief description of the goals of affirmative action for women, participants are asked to evaluate the effects of these programs on the situation of men by means of 6 items: 3 statements and 3 associated evaluative questions.
Sample items	1) General Attitudes toward Affirmative Action - If there are no affirmative action programs helping women in employment, they will continue to be unfairly treated. - After years of discrimination, it is only fair to set up special programs to make sure that women are given fair and equitable treatment. - All in all, do you favor the implementation of affirmative action programs for women in industries? 2) Impact on Men's Collective Interest - These programs disadvantage men compared to women in terms of their chances of getting a job. - These programs disadvantage men, compared with women, in terms of their chances for obtaining a promotion. Each statement is followed by a question asking participants whether they are satisfied with the implied situation.
Appropriate for whom (i.e. which population/s)	Adults

Sexism & Sex Discrimination

TITLE OF MEASURE	GENERAL ATTITUDES TOWARD AFFIRMATIVE ACTION (AA) AND MEN'S COLLECTIVE INTEREST (CI)

Translations & cultural adaptations available	English and French versions available Women and racial/ethnic minority versions available
How developed	Items were developed by the study authors.
Psychometric properties	*see below*

STUDY SAMPLES

Participants		Study 1	Study 2
Sample Size		n = 130	n = 149
Description		Students	Workers
Age	Range	18-43	29-60
	Mean	21.6	41.5
Gender	Male	100%	100%
Race/Ethnicity		Not reported	Not reported

VALIDITY

Concurrent Validity

	Men's Collective Interest Scale		Neosexism Scale	
	Study 1	Study 2	Study 1	Study 2
Scale	R =	r =	r =	r =
AA	-.48***	-.33***	-.58***	-.36***
CI			.50***	.18*

$*p < .05; ***p < .001$

RELIABILITY
Internal Consistency

Variable	Study 1	Study 2
Version	women	women & minority
Language	English & French	French
General Attitudes (AA)	α = .81	α = .86
Men's Collective Interest	α = .81	α = .67

Comments	• Samples for both validation studies were all male, and the ethnic/racial make-up of the sample was not reported. It would be useful to assess its validity and reliability for women and for multiple ethnic/racial groups.

Sexism & Sex Discrimination

TITLE OF MEASURE   GENERAL ATTITUDES TOWARD AFFIRMATIVE ACTION (AA) AND MEN'S COLLECTIVE INTEREST (CI)

- Since the scale was developed in Canada, some items may not translate to the situation in other countries (particularly given the wide range of approaches to affirmative action).

Bibliography (studies that have used the measure)

Contact Information

Francine Tougas
School of Psychology
University of Ottawa
Ottawa, Ontario KIN 6N5, Canada

e-mail: ftougas@uottawa.ca

Sexism & Sex Discrimination

TITLE OF MEASURE	NEOSEXISM SCALE
Source/Primary reference	Tougas, F., Brown, R., Beaton, A. M., & Joly, S. (1995). Neosexism: Plus a change, plus c'est pareil. *Personality and Social Psychology Bulletin, 21*(8), 842-891.
Construct measured	Neosexism defined as the "manifestation of a conflict between egalitarian values and residual negative feelings toward women."
Brief description	This instrument consists of 11 items based on the tenets of modern racism (McConahay, 1986). The rating scale ranges from 1 to 7 where 1 = total disagreement and 7 = total agreement.
Sample items	▪ Women will make more progress by being patient and not pushing too hard for change. ▪ Discrimination against women in the labor force is no longer a problem. ▪ In order to not appear sexist, many men are inclined to overcompensate women.
Appropriate for whom (i.e. which population/s)	Adults
Translations & cultural adaptations available	English and French versions available
How developed	The authors developed a number of items specifically for this scale and adapted items from covert racism scales, due to their relevance to the situation of women. An exploratory factor analysis did not show a definite structure so all the items were pooled.
Psychometric properties	*STUDY SAMPLES*

Participants		Study 1	Study 2
Sample Size		n = 130	n = 149
Description		Students	Workers
Age	Range	18-43	29-60
	Mean	21.6	41.5
Gender	Male	100%	100%
Race/Ethnicity		Not reported	Not reported

Sexism & Sex Discrimination

TITLE OF MEASURE	NEOSEXISM SCALE

VALIDITY

Concurrent Validity

Scale	Affirmative Action	
	Study 1 $r =$	Study 2 $r =$
Neosexism Scale	-.58***	-.36***

***$p < .001$

RELIABILITY

Internal Consistency & Test-Retest Reliability

Variable	Study 1	Study 2
Language	English & French	French
Cronbach's $\alpha =$	.78	.76
Test-retest $r =$	.84**	-

**$p < .01$

Comments	▪ Samples for both validation studies were all male, and the ethnic/racial make-up of the sample was not reported. It would be useful to assess its validity and reliability for women and for multiple ethnic/racial groups.
	▪ Neosexism is an interesting construct that assesses support for public policies designed to support women, while most sexism measures look at prejudicial attitudes or discriminatory behavior based on gender.
	▪ When compared with the Modern Sexism Scale, the Neosexism Scale was found to have better internal reliability and exhibited stronger gender differences (Campbell et al., 1997).
Bibliography (studies that have used the measure)	Campbell, B., Schellenberg, E. G., & Senn, C. (1997). Evaluating measures of contemporary sexism. *Psychology of Women Quarterly, 21*(1), 89-102. Masser, B., & Abrams, D. (1999). Contemporary sexism: The relationships among hostility, benevolence, and neosexism. *Psychology of Women Quarterly, 23*, 503-517.
Contact Information	Francine Tougas School of Psychology University of Ottawa Ottawa, Ontario K1N 6N5, Canada e-mail: ftougas@uottawa.ca

Sexual Harassment

Sexual Harassment

TITLE OF MEASURE	RESPONSES TO SEXUAL HARASSMENT AND SATISFACTION WITH THE OUTCOME
Source/Primary reference	Bingham, S. G., & Scherer, L. L. (1993). Factors associated with responses to sexual harassment and satisfaction with outcome. *Sex Roles, 29*(3/4), 239 -269.
Constructs measured	Three constructs: 1. Work climate regarding sexual harassment 2. Responses to sexual harassment 3. Satisfaction with the outcome
Brief description	*Work Climate Regarding Sexual Harassment* is assessed with a 3-item instrument, rated on a 5-point scale from "strongly agree" to "strongly disagree." *Responses to Sexual Harassment* are assessed with two checklists. First there is a five-item checklist of general responses to a sexual harassment situation. Second, there is a communications strategy checklist to be filled out only by those who reported that they talked to the harasser. *Satisfaction with Outcome* is a one-item measure about the victim's satisfaction with the outcome, rated on a 4-point scale from "definitely not" to "yes, definitely."
Sample items	Work Climate Regarding Sexual Harassment: - Sexual harassment is clearly discouraged by my supervisors and co-workers. - People in my department ignore sexual harassment. - The general attitude toward sexual communication in my department actually encourages sexual harassment. ***Responses to Sexual Harassment*** - Response checklist included: filing a formal complaint, informally talking to an external authority (e.g., ombudsperson, affirmative action officer), informally talking to an internal authority (e.g., supervisor, chair of the department), talking to co-workers, talking to friends or family members, talking to the harasser. - Communication strategies checklist (for the sub-sample who talked to the harasser) included: indirect communication strategies (e.g., ignoring or joking about the person's behavior, hinting that the behavior was unwelcome), assertive communication strategies (e.g., asking the person to stop, stating objections to the behavior), and

Sexual Harassment

TITLE OF MEASURE | RESPONSES TO SEXUAL HARASSMENT AND SATISFACTION WITH THE OUTCOME

aggressive communication strategies (e.g., expressing anger and hostility, using threats to get the person to stop).

Satisfaction with Outcome

- Did the situation involving unwanted sexual communication get resolved to your satisfaction?

Appropriate for whom (i.e. which population/s)	Adults
Translations & cultural adaptations available	None known
How developed	The study was a part of a larger, ongoing effort by a medium-sized Midwestern university to reduce, if not eliminate, sexual harassment on campus. The questionnaire domains and dimensions were extracted from the literature, and specific items were written by the study authors.

Psychometric properties

<u>STUDY SAMPLE</u>

Participants		Demographics
Sample Size		$n = 105$
Description		Employees of a Midwestern U.S. university who had reported experiencing unwanted sexual attention from a faculty, staff, or student member of the university.
Ethnicity	Caucasian	94 (89.5%)
	Multiethnic/ multicultural	6 (5.7%)
	Not specified	5 (4.8%)
Gender	Female	68 (65%)
	Male	37 (35%)
Position	Staff members	51 (49%)
	Faculty members	47 (45%)
	Not specified	7 (4%)

<u>RELIABILITY</u>

Internal consistency

The Cronbach α reliability for the Work Climate subscale is .78.

Comments

- The measures are brief, user-friendly approaches to assessing elements of the work climate and responses relevant to sexual harassment situations.

Sexual Harassment

TITLE OF MEASURE — RESPONSES TO SEXUAL HARASSMENT AND SATISFACTION WITH THE OUTCOME

- There is limited information available about psychometric properties.

- The full study uses multiple methods: checklists, open-ended questions, Likert-type items, and trained coders to classify certain responses into different categories.

- The authors were concerned about the inflation of alpha and type I errors that result when a series of univariate tests are conducted.

Bibliography (studies that have used the measure)

Contact Information

Shereen G. Bingham, Associate Professor
Communication Faculty and Women's Studies Faculty
Arts and Science Hall, Rm140
University of Nebraska at Omaha
Omaha, NE 68182-0112
Tel: 402-554-4857

e-mail: sbingham@mail.unomaha.edu

http://avalon.unomaha.edu/wwwcomm/Faculty%20pages/sbingham.html

Lisa L. Scherer, Associate Professor
Psychology Faculty
Arts and Sciences Hall, Rm 347-J
University of Nebraska at Omaha
Omaha, NE 68182
Tel: 402-554-2698

e-mail: lscherer@mail.unomaha.edu

Sexual Harassment

TITLE OF MEASURE	SEXUAL EXPERIENCES QUESTIONNAIRE - LATINAS (SEQ-L)
Source/Primary reference	Cortina, L. M. (2001). Assessing sexual harassment among Latinas: Development of an instrument. *Cultural Diversity and Ethnic Minority Psychology, 7*(2), 164-181.
Construct measured	Experiences of sexual harassment among Latinas
Brief description	A variation of the Sexual Experiences Questionnaire (SEQ) (Fitzgerald et al., 1995) developed for Latinas. The scale has 20 items rated on a 5-point scale where responses range from 1 = never to 5 = most of the time. There are three subscales: 1. Sexist hostility 2. Sexual hostility 3. Unwanted sexual attention
Sample items	<ul><li>Told jokes or stories that described women IN GENERAL negatively?</li><li>Said things to insult LATINA women specifically (for example saying that Latinas are "hot-blooded" and "loose")?</li><li>Made you uncomfortable by staring at you (for example, looking at you too long, or looking at your breasts)?</li><li>Gave you sexual attention that you did not want?</li><li>Made you uncomfortable by standing too close?</li></ul>
Appropriate for whom (i.e. which population/s)	Latina women
Translations & cultural adaptations available	English and Spanish versions available
How developed	A first study was conducted to understand the experience of sexual harassment of Latinas and to guide the development of a culturally relevant measure. Focus groups were organized with Latina women students in an adult education program. As a result, six items were added to the dimensions of SEQ as developed by Fitzgerald and colleagues. Two of these items referred to specific verbal behaviors and four referred to nonverbal behaviors. Another five items were added to measure sexual racism. In order to avoid the influence of ethnicity and gender in

the responses, these questions about Latinas were paired with the same question asked for women in general. As a result, 16 new questions developed for this study were added to the existing SEQ.

In a second study to validate the scale, the researchers had a sample of women from vocational training programs in San Diego and Chicago. The results showed that two of the newly developed items had a low variance and were thus eliminated. Two other items were dropped because participants could interpret them in more than one way, such that the behavior did not necessarily qualify as sexual harassment. Therefore, not enough items were left to consider a sexual-racism factor.

In the end, nine items were developed for this particular scale and eleven were taken from SEQ, resulting in a 20-item Sexual Experiences Questionnaire for Latinas.

Psychometric properties

STUDY SAMPLE

Participants		Study 1
Sample Size		$N = 462$
Description		Latinas from public adult schools, job training centers, or a "swap meet" in the San Diego area and from a public adult school in a Chicago suburb
Age	Range	18-55+
Education	Graduate School	$n = 22$
	College	$n = 74$
	High School+	$n = 192$
	Less than High School Diploma	$n = 167$
Marital Status	Single	$n = 215$
	Married/Living with Partner	$n = 178$
	Widowed	$n = 7$
	Separated/Divorced	$n = 60$

RELIABILITY

The α reliability for the full scale is .96 for both the English and Spanish versions.

Sexual Harassment

TITLE OF MEASURE   SEXUAL EXPERIENCES QUESTIONNAIRE - LATINAS (SEQ-L)

Subscale	Spanish $\alpha =$	English $\alpha =$	Overall $\alpha =$
Sexist Hostility	.91	.88	.90
Sexual Hostility	.90	.89	.90
Unwanted Sexual Attention	.94	.96	.95

Comments

The scale builds on a well-validated scale and expends its usefulness with a new population.

- The author used a convenience sample, composed of adult education students. This approach limits reaching immigrants without legal status, very-low-income women, and professional workers.

- Most of the participants in the study were of Mexican origin: for expanded use, the scale should be validated with immigrant groups from other Latin American cultures.

Bibliography (studies that have used the measure)

Cortina, L. (2002). Contextualizing Latina experience of sexual harassment: Preliminary tests of a structural model. *Basic and Applied Social Psychology, 24*(4), 295-311.

Contact Information

Lilia M. Cortina
Department of Psychology
University of Michigan
525 East University
Ann Arbor, MI 48109-1109, USA

Sexual Harassment

TITLE OF MEASURE	SEXUAL EXPERIENCE QUESTIONNAIRE (SEQ-W)
Source/Primary reference	Fitzgerald, L. F., Gelfand, M. J., & Drasgow, F. (1995). Measuring sexual harassment: Theoretical and psychometric advances. *Basic and Applied Social Psychology, 17*(4), 425-445.
Construct measured	Experiences of sexual harassment in the workplace
Brief description	An original SEQ scale consisted of 28 items. The revised SEQ-W scale contains 20 items and is made up of 3 subscales and a separate item retained as an individual category to measure the participant's subjective perceptions of sexual harassment. The rating scale ranges from 0 = never to 4 = many times. The three subscales measure: 1. Gender Harassment 2. Unwanted Sexual Attention 3. Sexual Coercion
Sample items	▪ Have you ever been in a situation where a supervisor or coworker habitually told suggestive stories or offensive jokes? ▪ Have you ever been in a situation where your coworker made unwanted attempts to stroke or fondle you? ▪ Have you ever been in a situation where you felt that you were being subtly bribed with some sort of reward to engage in a sexual behavior with a coworker?
Appropriate for whom (i.e. which population/s)	Employed women
Translations & cultural adaptations available	A version designed for students is available as SEQ-E. A version designed for Latinas, developed by Cortina (2001) is available as SEQ-L (see previous entry).
How developed	The first version of the scale was developed in 1988 and showed good psychometric properties for research purposes. However, the authors saw the need to improve the scale in order to distinguish between the type and the severity of harassment, change the dimensional structure of the scale, and use more sensitive wording. The authors developed new items and revised the previous ones to fit the three-dimensional model. The SEQ-W was first examined in a large utility company.

Sexual Harassment

TITLE OF MEASURE   *SEXUAL EXPERIENCE QUESTIONNAIRE (SEQ-W)*

Psychometric properties

<u>STUDY SAMPLES</u>

Participants	Regulated Utility	Agricultural Organization	Midwestern University
Sample Size	$N = 448$	$N = 410$	$N = 299$
Description	Employees of a West Coast public utility company	Employees of agribusiness factory sites	Employees of a Midwestern university
Gender	100% women	100% women	100% women
Race/Ethnicity	Not reported	Not reported	Not reported

<u>VALIDITY</u>

Concurrent Validity

Correlations of SEQ-W with other measures, in employees of a regulated utility ($n = 448$), agribusiness factory sites ($n = 410$), and university ($n = 299$).

Measure	Regulated Utility	Agricultural Organization	Midwestern University
OTSHI	.45**	.23**	.40**
SUPSAT	-.18**	-.23**	-.36**
COWSAT	-.29**	-.16**	-.26**
WKSAT	-.09	-.11*	-----
JOBWITH	.11*	.20**	-----
WKWITH	.19**	.32**	.19**
HELSAT	-.03	-.07	-----
HELCOND	-.09	-----	-----
LIFESAT	-.08	-.16**	-.20**
PTSD	.17**	-----	.19**
DISTRESS	.16**	.11*	-----
SIG	.05	.17**	.20**
EXTCOM	.14*	-----	-----

$*p < .05; **p < .01$

- OTSHI = Organizational Tolerance for Sexual Harassment Inventory
- SEQ = Sexual Experiences Questionnaire
- SUPSAT = JDI Satisfaction with Supervision
- COWSAT = JDI Satisfaction with Coworkers
- WKSAT = JDI Satisfaction with Work
- JOBWITH = Job Withdrawal
- WKWITH = Work Withdrawal
- HELSAT = RDI Health Satisfaction
- HELCOND = Health Conditions Index

Sexual Harassment

TITLE OF MEASURE — SEXUAL EXPERIENCE QUESTIONNAIRE (SEQ-W)

- LIFESAT = Life Satisfaction Scale + Faces Scale
- PTSD = Crime-Related Post-Traumatic Stress Disorder
- DISTRESS = MHI Distress
- SIG = Stress in General
- EXTCOM = Extrinsic Organizational Commitment

RELIABILITY

Reliabilities for the *SEQ-W* in 3 samples

Subscale	Regulated Utility $\alpha =$	Agricultural Organization $\alpha =$	Midwestern University $\alpha =$
Gender Harassment	.81	.78	.72
Unwanted Sexual Attention	.82	.80	.67
Sexual Coercion	.41	.92	.49
Overall	.86	.88	.78

Comments

- This is the best validated scale for assessing experiences of sexual harassment and has been used by a wide range of researchers.

- One disadvantage of the scale is its length, and thus some other researchers seem to adopt alternative approaches that are less detailed.

- There are demonstrated relationships to both psychological states (i.e., anxiety and depression) and physical health (Fitzgerald et al., 1997).

- There are versions designed specifically for work settings and alternative versions developed for academic settings.

- Some work has been done to adapt the scale for use with Latinas (see Cortina entry). However, the ethnic/racial make-up of the sample was not reported in the articles we reviewed that describe the development of the initial scale.

Bibliography (studies that have used the measure)

Fitzgerald, L. F., Drasgow, F., Hulin, C. L., Gelfand, M. J., & Magley, V. (1997). Antecedents and consequences of sexual harassment in organizations: A test of an integrated model. *Journal of Applied Psychology, 28*(4), 578-589.

Fitzgerald, L. F., Shullman, S., Bailey, N., Richards, M., Swecker, J., Gold, A., Ormerod, A. J., & Weitzman, L. (1988). The incidence and dimensions of sexual harassment in academia and the workplace. *Journal of Vocational Behavior, 32*, 152-175.

Sexual Harassment

TITLE OF MEASURE	**SEXUAL EXPERIENCE QUESTIONNAIRE (SEQ-W)**
	Gelfand, M. J., Fitzgerald, L. F., & Drasgow, F. (1995). The structure of sexual harassment: A confirmatory analysis across cultures and settings. *Journal of Vocational Behavior, 47*, 164-177.
Contact Information	Louise F. Fitzgerald
Department of Psychology
University of Illinois at Urbana-Champaign
603 East Daniel St.
Champaign, IL 61820, USA

No cost from the author. |

Sexual Harassment

TITLE OF MEASURE	ORGANIZATIONAL TOLERANCE FOR SEXUAL HARASSMENT INVENTORY (OTSHI)
Source/Primary reference	Hulin, C., Fitzgerald, L., & Drasgow, F. (1996). Organizational influences on sexual harassment, in M. Stockdale (Ed.), *Sexual Harassment in the Workplace: Perspectives, frontiers and response strategies*. Thousand Oaks, CA: Sage.
Construct measured	Perceptions of likelihood of organizational reactions to various forms of harassment
Brief description	The OTSHI instrument consists of six brief vignettes, in which the characteristics of a male harasser (supervisor, or coworker) are crossed with each of three types of sexual harassment: gender harassment, sexual coercion, unwanted sexual attention. After each vignette, respondents are asked to make three assessments using 5-point rating scales (18 items total): 1. The degree of risk to a female victim if she reported such an incident 2. The likelihood that her allegations would be taken seriously by the organization 3. The likelihood that the harasser would receive meaningful sanctions
Sample items	Gender Harassment x Supervisor Scenario ▪ A supervisor in your department makes reference to "incompetent women trying to do jobs they were never intended to do and taking jobs away from better qualified workers." He makes all women in the department feel incompetent and unwanted. Unwanted Sexual Attention x Coworkers ▪ An employee in your department continues to pressure the women in the department to go out with him after they have made it clear that they are not interested.
Appropriate for whom (i.e. which population/s)	Adult women and men
Translations & cultural adaptations available	None known
How developed	Items for the scale were written based on a facet analysis of harassing incidents that suggested two facets: organizational role of the harasser and type of harassing behavior. The facet analysis generated a six-cell design

Sexual Harassment

TITLE OF MEASURE   ORGANIZATIONAL TOLERANCE FOR SEXUAL HARASSMENT INVENTORY (OTSHI)

that crosses two harasser roles and three types of harassing behavior. The authors included 36 items in the scale to be piloted with a sample of 263 graduate students. The internal consistency of the scale was .96, and since none of the items accounted for a unique variance, the authors decided to shorten the scale by eliminating one item in each cell, leaving a final version with 18 items.

Psychometric properties

STUDY SAMPLE

Participants		Demographics
Sample Size		$n = 1,156$
Gender	Female	$n = 459$
	Male	$n = 697$
	Race/Ethnicity	Not reported

VALIDITY

Construct Validity

				Measurement model factor loading		
Scale	$M =$	$SD =$	$\alpha =$	1	2	3
OTSHI	2.25	0.86	.96	.89	.92	.94

RELIABILITY

	Reliability	
	Female	Male
Subscale	$\alpha =$	$\alpha =$
Risk	.94	.89
Serious	.94	.91
Action	.93	.91

Comments

- This is an innovative approach to measuring an aspect of the organizational context.

- The instrument requires a fairly literate study population.

- A simpler but less rigorous approach to assessing organizational tolerance is described in Hesson-McInnis and Fitzgerald (1997).

- The ethnic/racial make-up of the sample was not reported. It would be useful to assess the scale's validity and reliability for multiple ethnic/racial groups.

Sexual Harassment

TITLE OF MEASURE	ORGANIZATIONAL TOLERANCE FOR SEXUAL HARASSMENT INVENTORY (OTSHI)
Bibliography (studies that have used the measure)	Fitzgerald, L. F., Drasgow, F., Hulin, C. L., Gelfand, M. J., & Magley, V. J. (1997). Antecedents and consequences of sexual harassment in organizations. *Journal of Applied Psychology, 82*(4), 578-590. Hesson-McInnis, M., & Fitzgerald, L. (1997). Sexual harassment: A preliminary test of an integrated model. *Journal of Applied Social Psychology, 27*(10), 877-901.
Contact Information	Louise F. Fitzgerald Department of Psychology University of Illinois at Urbana-Champaign 603 East Daniel Street Champaign, IL 61820, USA No cost from the author.

Sexual Harassment

TITLE OF MEASURE	SEXUAL HARASSMENT INVENTORY (SHI)
Source/Primary reference	Murdoch, M., & McGovern, P. G. (1998). Measuring sexual harassment: Development and validation of the Sexual Harassment Inventory. *Violence and Victims, 13*(3), 203 - 216.
Construct measured	Sexual harassment
Brief description	The scale includes a list of 20 behaviors of a sexual nature and one open-ended question. Item responses are "yes" or "no." The items can also be weighted by severity. After the factor analyses, three SHI subscales emerged: (i) Hostile Environment, (ii) Quid Pro Quo, and (iii) Criminal Sexual Misconduct.
Sample items	▪ People with whom I worked made sexual jokes that made me feel uncomfortable. ▪ Coworkers made sexual comments about my body. ▪ I was offered favorable assignments in exchange for sex with my supervisor (or, in the military version, commanding officer). ▪ Some of the people I worked with leered at me in a sexual way. ▪ The people I worked with made catcalls or sexual remarks when I walked by.
Appropriate for whom (i.e. which population/s)	Adults
Translations & cultural adaptations available	Active-duty and veteran military populations.
How developed	Items for inclusion in the SHI were gathered from three focus groups, literature reviews, and court cases involving sexual harassment. A pilot study with 80 male and female students was conducted to solicit feedback to improve wording and clarity. Then the SHI was modified to apply to the military environment.
Psychometric properties	*STUDY SAMPLE* Study 1: The first study sample included female veterans (n = 333) who had obtained medical care at the MVAMC between March 1992 and March 1993. Information about race/ethnicity is not included. Respondents were asked to indicate whether they had experienced any of the behaviors listed in the SHI while they were in the military.

| TITLE OF MEASURE | SEXUAL HARASSMENT INVENTORY (SHI) |

Study 2: A second study involved anonymous surveys of Equal Employment Opportunity (EEO) officers (n = 160) stationed at Veterans Affairs hospitals throughout the U.S. Officers were asked to provide severity weights to each item on the SHI. The ranks ranged from 1 (least severe) to 20 (most severe). 64% of the respondents were female. Race/ethnicity is not reported.

The author reports that reliability and factor structure is similar for both white persons and non-white persons.

VALIDITY

Content Validity

Items for inclusion in the SHI were gathered from three focus groups, literature reviews, and court cases involving sexual harassment. The focus group discussion were held with:

- physicians employed at the Minneapolis Veterans Affairs Medical Center (MVAMC) (n = 3),

- sitting members of the MVAMC Sexual Trauma Treatment Team (n = 8), and

- a convenience sample of women who accompanied their husbands to clinic visits at the MVAMC (n = 10).

In a comprehensive sample of Tri-Care and CHAMPUS enrollees who received care at a Midwestern VA medical facility (n = 293 men and 237 women), 85% agreed or strongly agreed that the SHI measured their most important experiences with unwanted sexual attention while they were in the service.

Construct Validity

Factor analysis was conducted using oblique (Oblimin) and orthogonal (varimax) rotations. Items were considered to be salient to their respective factors if factor loadings were greater than or equal to .40. Items were considered to be exclusive to the factor if the difference between their salient factor loading and loadings on other factors was greater than .11.

Three factors accounted for 57% of the variance in the model. Whereas the first two factors appeared to correspond to the latent variables "hostile environment" and "quid pro quo," the third and

Sexual Harassment

TITLE OF MEASURE   SEXUAL HARASSMENT INVENTORY (SHI)

the smallest factor seemed to be related to the most serious forms of sexual harassment, such as rape and attempted rape (called in this study "criminal sexual misconduct"). All SHI items were salient to their factors. With one exception, all were exclusive.

In military populations, the severity-weighted SHI has been shown to correlate in the expected direction with symptoms of posttraumatic stress disorder, depression, anxiety, and somatization; with work, role, social, and physical functioning; and with other trauma experiences.

Concurrent Validity

The EEO officers' severity weightings did not differ significantly by gender or by full-time or part-time status. Kruskal's stress index was .088, and r^2 was .98, indicating that less than 1% of the variance of the model was due to error, and 98% of the variance was explained by a unidimensional model. Therefore, it was highly unlikely that attributes other than the severity accounted for the manner in which respondents ranked the scale items.

The SHI correlates .88 with a modified version of the Sexual Experiences Questionnaire.

Criterion-Related Validity

Three items from the "criminal sexual misconduct" scale plus the open-ended question had sensitivity of 90%, specificity of 100%, and overall accuracy of 91% in identifying in-service sexual assault in a small sample ($n = 11$) of women veterans who filed disability claims for posttraumatic stress disorder. Objective, second-party documentation from the veterans' claims files (e.g., police reports, hospital reports, testimonials from friends who remembered being told about the assault at the time it occurred) served as the gold standard.

In a randomly selected sample of women veterans applying for posttraumatic stress disorder disability benefits ($n = 1,682$), severity-weighted SHI scores greater than 20.01 were associated with 2-fold greater odds of meeting survey criterion for posttraumatic stress disorder compared to women with lower SHI scores, even after accounting for other adult trauma experiences.

Sexual Harassment

TITLE OF MEASURE	SEXUAL HARASSMENT INVENTORY (SHI)

<table>
<tr><td></td><td>

RELIABILITY

Internal Consistency

The Cronbach α reliability for the full SHI scale ranges from .90 to .95

The Cronbach α reliabilities of three SHI subscales

SHI subscale	α =
Hostile environment	.89
Quid pro quo	.86
Criminal sexual misconduct	.86

</td></tr>
<tr><td>

Comments

</td><td>

- The SHI content domain is comprehensive. Its internal consistency reliability is high, and it has evidence of factorial validity.

- The behaviors measured in the SHI can be weighted according to relative severity, allowing investigators to examine dose-response relationships and threshold between sexual harassment and various health outcomes - especially mental health outcomes such as anxiety, depression, alcohol misuse, or PTSD.

- Other uses of the severity-weighted SHI might include associations between sexual harassment and measures of well-being such as social adjustment or quality of life.

- The instrument can be easily adapted for use in a general working population by replacing "commanding officer" with "supervisor."

- The authors allow other investigators to freely reproduce and use the SHI as long as the article and journal are referenced.

</td></tr>
<tr><td>

Bibliography (studies that have used the measure)

</td><td>

Murdoch, M., Hodges, J., Cowper, D., Fortier, L., & vanRyn, M. (2003). Racial disparities in VA service connection for posttraumatic stress disorder disability. *Medical Care, 41*(4), 536-549.

Murdoch, M., Polusny, M. A., Hodges, J., & O'Brien, N. (2004). Prevalence of in-service and post-service sexual assault among combat and noncombat veterans applying for Department of Veterans Affairs posttraumatic stress disorder disability benefits. *Military Medicine, 169*, 392-395.

Halek, K., Murdoch, M., & Fortier, L. (2005). Spontaneous reports of emotional upset and health care utilization among veterans with posttraumatic stress disorder (PTSD) after receiving a potentially upsetting survey. *Journal of American Orthopsychiatry, 75*(1):142-151.

</td></tr>
</table>

Sexual Harassment

TITLE OF MEASURE	SEXUAL HARASSMENT INVENTORY (SHI)

Murdoch, M., Polusny, M. A., Hodges, J., & Cowper, D. (2006). The association between in-service sexual harassment and posttraumatic stress disorder among Department of Veterans Affairs disability applicants. *Military Medicine 171*(8):166-173.

Murdoch, M., Pryor, J., Hodges, J., Gackstetter, G. D., Cowper, D., & O'Brien, N. Findings from VA HSR&D Project #IIR-96-014, "Antecedents and Consequences of Military Sexual Trauma" Final Report.

Contact Information

Maureen Murdoch
Associate Professor of Medicine
Center for Chronic Disease Outcomes Research
(a VA HSR&D Center of Excellence) and
Section of General Internal Medicine

Office/Campus address:
Minneapolis Veterans Affairs Medical Center
One Veterans Dr. (111-0)
Minneapolis, MN 55416, USA

e-mail: Maureen.Murdoch@med.va.gov

Work Family/Work-Life Measures

Work Family/Work-Life Measures

TITLE OF MEASURE	**WORK-HOME CONFLICT**
Source/Primary reference	Bacharach, S. B., Bamberger, P., & Conley S. (1991). Work-home conflict among nurses and engineers: Mediating the impact of role stress on burnout and satisfaction at work. *Journal of Organizational Behavior, 12*(1), 39-53.
Construct measured	Interrole conflict in which the role pressures from work and family (home) domains feel mutually incompatible
Brief description	This scale is designed to tap the degree to which the job impacts upon and/or disrupts the individual's life at home. It consists of 4 items which are rated in terms of frequency on a scale of 1 = seldom or never to 4 = almost always.
Sample items	▪ Do the demands of work interfere with your home, family or social life? ▪ Does the time you spend at work detract from your family or social life? ▪ Does your work have disadvantages for your family or social life? ▪ Do you not seem to have enough time for your family or social life?
Appropriate for whom (i.e. which population/s)	Working adults
Translations & cultural adaptations available	None known
How developed	It is a four-item scale based on that of Holahan and Gilbert (1979).
Psychometric properties	<u>STUDY SAMPLE</u>

Participants	Nurses	Civil Engineers
Sample Size	$n = 215$	$n = 430$
Description	Employees of a large state in the Northeast	
Gender	Not reported	Not reported
Race/Ethnicity	Not reported	Not reported

Work Family/Work-Life Measures

TITLE OF MEASURE   WORK-HOME CONFLICT

VALIDITY

Construct Validity

Scores on the Work-Home Conflict Scale were positively correlated with general role conflict and role overload and negatively correlated with job satisfaction.

RELIABILITY

Internal Consistency

Scale	Nurses $\alpha =$	Civil Engineers $\alpha =$
Work-Home Conflict	.87	.77

Comments

- This scale is sensitive to a broad range of concerns and works for both married and unmarried employees.

- The fact that no gender or race/ethnicity demographics are presented is problematic. It would be useful to know its validity and reliability for multiple groups.

Bibliography (studies that have used the measure)

Contact information

Samuel Bacharach
ILR Organizational Behavior
200 ILR Ext. Bldg.
Cornell University
Ithaca, NY 14853, USA

Tel: 607-255-2772

e-mail: sb22@cornell.edu

Work Family/Work-Life Measures

TITLE OF MEASURE: *PARENTAL AFTER-SCHOOL STRESS (PASS)*

Source/Primary reference	Barnett, R. C., & Gareis, K. C. (under review). Parental after-school stress and psychological well-being. Manuscript submitted for publication in *Journal of Marriage and Family*.
Construct measured	Degree to which employed parents are concerned about the welfare of their school-aged children during the after-school hours
Brief description	The measure contains 10 items. Respondents indicate their level of concern about their target child's after-school arrangements in a variety of domains including safety, travel, productive use of time, and dependability, among others. Items are rated on a 4-point scale from 1 = not at all to 4 = extremely.
Sample items	▪ How much do you worry about your school-aged child's travel to and from (his/her) after-school arrangements? ▪ How much do you worry that your school-aged child's after-school arrangements will fall through? ▪ How much do you worry about whether your school-aged child is spending (his/her) after-school time productively?
Appropriate for whom (i.e. which population/s)	Employed parents of school-aged (i.e., K-12) children, whether or not those children are in formal after-school arrangements
Translations & cultural adaptations available	None known
How developed	Items were generated by the researchers and further refined through two stages of pilot testing with employees at all levels of a Boston-area utility company. After the draft measure was developed, focus groups were convened with mothers and fathers of children in grades K through 12 for a general discussion of their issues with after-school arrangements; afterwards, participants were asked to give feedback on the draft measure which was then used to refine it. In the next stage, 59 employees at the same company completed mail surveys; based on these findings, the authors further refined the PASS measure.
Psychometric properties	*STUDY SAMPLES* The revised measure has been administered to and validated in (1) a small sample of employees who have school-aged children and who work at a Boston-area consumer goods company, (2) a small sample of employees who have school-aged children and who work at a North Carolina software company, and (3) a larger sample of employees in six states who have school-aged children and who work for a large financial

Work Family/Work-Life Measures

TITLE OF MEASURE   PARENTAL AFTER-SCHOOL STRESS (PASS)

services company. The measure was administered as a web-based survey to Samples 1 and 2 and as a mailed survey to Sample 3. The authors are currently administering the measure to a community sample of parents with school-aged children in three family types: dual-earner couples, single-breadwinner couples, and employed single parents.

Participants		Sample 1	Sample 2	Sample 3
Sample Size		36	36	243
Age	Range	24-47	32-55	25-59
	Mean (SD)	38.0 (5.7)	41.3 (5.5)	39.2 (6.3)
Gender	Female	31	31	205
	Male	5	5	38

VALIDITY

Construct Validity

The measure of parental after-school stress (PASS) is related to other variables in predicted ways. For example, in Sample 3, the authors found that parents whose jobs are less flexible and whose children spend more time unsupervised by an adult after school report significantly higher levels of PASS, and that parents with high PASS report significantly higher levels of job disruptions and significantly lower levels of psychological well-being (Barnett, 2003; Barnett & Gareis, under review).

RELIABILITY

Internal Consistency

Sample	Cronbach's α
1. Employees who have school-aged children and who work at a Boston-area consumer goods company	.76
2. Employees who have school-aged children and who work at a North Carolina software company	.82
3. Employees who have school-aged children and who work at a large financial services company	.87

Comments

- More than one-third (37.2%) of the labor force consists of parents of minor children, the majority of those children are of school age. However, most parents have work schedules that prevent them from being home when their children get out of school, leaving a substantial gap between the time the school day ends and the time most parents get home from work.

Work Family/Work-Life Measures

TITLE OF MEASURE: PARENTAL AFTER-SCHOOL STRESS (PASS)

	- Note that one manuscript on this measure is currently being revised for resubmission to a peer-reviewed journal, and a second is in preparation after being invited for a special issue of a peer-reviewed journal.
Bibliography (studies that have used the measure)	Barnett, R. C., & Gareis, K. C. (2004, July/August). Parental after-school stress, psychological distress, and job performance. Paper presented at the annual meeting of the American Psychological Association, Honolulu, HI. Barnett, R. C. (2003, June). *Community:* The missing link in work-family research. Paper presented at the Workforce/Workplace Mismatch: Work, Family, Health, and Well-being conference, Washington, DC. Barnett, R. C., & Gareis, K. C. (2005) Predictors and consequences of parental after-school stress. Manuscript in preparation for special issue of *American Behavioral Scientist*.
Contact Information	Rosalind Chait Barnett Community, Families & Work Program Brandeis University Women's Studies Research Center Mailstop 079, 515 South Street Waltham, MA 02453-2720, USA Tel: 781-736-2287 e-mail: rbarnett@brandeis.edu

Work Family/Work-Life Measures

TITLE OF MEASURE	WORK SCHEDULE FIT
Source/Primary reference	Barnett, R. C., Gareis, K. C., & Brennan, R. T. (1999). Fit as a mediator of the relationship between work hours and burnout. *Journal of Occupational Health Psychology, 4*, 307-317.
Construct measured	Degree to which work schedule meets own and family needs
Brief description	The scale includes 11 items in three domains: 1. Fit of own schedule for oneself (self/self schedule fit) 2. Fit of own schedule for other family members; i.e., partner, children, elderly dependents (self/family schedule fit) 3. Fit of partner's schedule, if applicable, for all family members; i.e., self, partner, children, elderly dependents (partner/family schedule fit) Items are rated on a 7-point scale from 1 = extremely poorly to 7 = extremely well.
Sample items	Self/self schedule fit • Taking into account your current work hours and schedule, how well is your work arrangement working for you? Self/family schedule fit • Taking into account your current work hours and schedule, how well is your work arrangement working for your child(ren), if any? Partner/family schedule fit • Taking into account your partner's current work hours and schedule, how well is (his/her) work arrangement working for your elderly dependent(s), if any?
Appropriate for whom (i.e. which population/s)	People who are employed outside the home; especially relevant for workers with partners/families, but the self/self subscale can be used with any worker
Translations & cultural adaptations available	None known
How developed	Items were generated based on a review of the literature on work schedules and on the work-family interface. Workers are conceptualized as members of family systems who make and evaluate decisions about

	Work Family/Work-Life Measures
TITLE OF MEASURE	WORK SCHEDULE FIT

family members' work schedules based on consideration of the needs of all members of the family system. Work schedule fit is the extent to which workers have been able to optimize their work-family strategies, meeting their own and their family members' needs.

Psychometric properties

STUDY SAMPLES

The measure has been administered to and validated in (1) a sample of reduced-hours physicians and their employed partners, (2) a sample of full-time and reduced-hours female physicians and licensed practical nurses in dual-earner couples with children under high school age, and (3) a sample of day- and evening-shift registered nurses and their full-time employed partners with children between 8 and 14. We are currently administering the measure to a community sample of parents with school-aged (K-12) children in three family types: dual-earner couples, single-breadwinner couples, and employed single parents.

Participants		Sample 1	Sample 2	Sample 3
Sample Size		280	186	110
Age	Range	31-68	27-51	32-48
	Mean (SD)	42.6 (6.9)	40.1 (6.9)	43.3 (4.3)
Gender	Female	140	186	55
	Male	140		55
Race/ Ethnicity	Caucasian	92.5%	70.4%	94.5%
	African American	0.7%	7.5%	-
	Hispanic/Latino/Latina	2.5%	2.7%	-
	Asian	2.9%	18.8%	5.5%
	Other	1.4%	0.5%	-

VALIDITY

Construct Validity

The measure of work schedule fit is related to other variables in predicted ways. For example, fit is a better predictor of quality-of-life outcomes such as psychological distress, life satisfaction, burnout, job-role quality, and marital-role quality than is the number of work hours *per se* (Gareis & Barnett, 2001). In another study, the results of structural equation modeling show that the relationship between number of hours worked and burnout is mediated by work schedule fit in a sample of reduced-hours physicians; that is, at any level of work hours, physicians with poorer fit have higher levels of burnout at work (Barnett, Gareis, & Brennan, 1999).

Work Family/Work-Life Measures

TITLE OF MEASURE | WORK SCHEDULE FIT

RELIABILITY

Internal Consistency

Sample	Cronbach's α
1. Reduced-hours physicians and their employed partners	.70
2. Full-time and reduced-hours female physicians and licensed practical nurses in dual-earner couples with children under high school age	.70
3. Day- and evening-shift registered nurses and their full-time employed partners with children between 8 and 14	.77

Test-Retest Reliability

In Sample 1, a stability coefficient of $r = .83$ ($p = .000$) over an interval of one to three months indicates that the work schedule fit measure has high test-retest reliability.

Comments

Bibliography (3-5 recent studies that have used the measure)

Barnett, R. C., Gareis, K. C., & Brennan, R. T. (1999). Fit as a mediator of the relationship between work hours and burnout. *Journal of Occupational Health Psychology, 4*(4), 307-317.

Gareis, K. C., & Barnett, R. C. (2001, August). *Schedule fit and stress-related outcomes among women doctors with families*. Paper presented at the annual meeting of the American Psychological Association, San Francisco, CA.

Gareis, K. C., Barnett, R. C., & Brennan, R. T. (2003). Individual and crossover effects of work schedule fit: A within-couple analysis. *Journal of Marriage and Family, 65*(4), 1041-1054.

Contact Information

Rosalind Chait Barnett
Community, Families & Work Program
Brandeis University Women's Studies Research Center
Mailstop 079, 515 South Street
Waltham, MA 02453-2720, USA

Tel: 781-736-2287

e-mail: rbarnett@brandeis.edu

Work Family/Work-Life Measures

TITLE OF MEASURE	INFORMAL WORK ACCOMMODATIONS TO FAMILY (IWAF)
Source/Primary reference	Behson, S. J. (2002). Coping with family-to-work conflict: The role of informal work accommodations to family. *Journal of Occupational Health Psychology, 7*(4), 324-341.
Construct measured	Ways in which employees temporarily and informally adjust their usual work patterns in an attempt to balance their work and family responsibilities
Brief description	The scale includes 16 IWAF behaviors. Respondents are asked to rate how often they have exercised the behavior in question. The response alternatives range from 1 = never (about once a year or less) to 5 = very often (once or more per day). In addition, an open-ended question asks respondents to describe any other ways in which they have adjusted their work to address family concerns.
Sample items	Some employees adjust their typical work patterns in order to meet family responsibilities. Please think of the ways in which you may have done things differently at work in order to address family concerns. How often have you done each of the following things: • Arranging to leave work early in order to attend a family event. • Leaving work during the day but completing the work later that night (either at home or at the office). • Receiving family-related phone calls while at work. • Phoning or e-mailing family members from work. • Having your children come in to work so you can keep an eye on them.
Appropriate for whom (i.e. which population/s)	Working adults
Translations & cultural adaptations available	None known
How developed	The scale items were developed based on a literature review, two pilot studies, and several focus groups.

Work Family/Work-Life Measures

TITLE OF MEASURE   INFORMAL WORK ACCOMMODATIONS TO FAMILY (IWAF)

Psychometric properties

<u>STUDY SAMPLE</u>

Participants		Sample 1	Sample 2
Sample Size		n = 141	n = 128
Description		Employees of 10 branches of a large Northeast telecommunication company	Two mid-sized private secular universities in the Northeast and one small private religious college in the Southeast
Gender	Female	54.1%	59.4%
	Male	45.9%	40.6%
Marital Status	Married	65.7%	46%
	Not married	34.3%	54%
Job Categories	Managerial/Administrative	50.5%	-
	Sales	29.0%	-
	Clerical	5.9%	-
	Other	4.7%	-

Sample 1: 51.9% of the respondents had at least one child less than 18 years of age living with them. Among respondents, 66.7% of their spouses were employed full-time.

Sample 2: 44% of the respondents had at least one child less than 18 years of age living with them. The average household income of the respondents ranged from $20,000 to $200,000 and their average tenure at their current employer ranged from 1 to 264 months (22 years). The second study was conducted to provide evidence of the convergent and discriminant validity of the IWAF.

<u>VALIDITY</u>

Content Validity

The IWAF items were based on literature review and results of the two pilot studies. In the first stage of pilot testing, a number of informal, semi-structured interviews with a convenience sample of working parents were conducted. In the second stage, several focus groups were conducted in two separate organizations. In total, 37 people participated in all focus groups. Within each focus group, participants (i) read a consent form, (ii) were asked to write down a list of the ways in which they did things differently at work to accommodate family-related matters, (iii) filled out the IWAF scale, (iv) discussed how well the items in the IWAF scale reflected the actions in their lists and were asked to critique the scale, (v) discussed general work-family issues, and (vi) were given a copy of the full questionnaire to fill out on their own and return. The focus group participants suggested some changes for the IWAF scale content.

Concurrent Validity

Correlations between the IWAF scale and other related measures were derived.

Measure	IWAF Scale
Family-to-Work Conflict (Netemeyer, 1996)	.22
Ways of Coping (Lazarus & Folkman, 1984) - Problem-Focused Coping - Seeking Social Support - Emotion-Focused Coping	 .09 .23 -.40
Parental Responsibility Index-Responsibility for Dependents Scale (Rothausen, 1999)	.22
Financial Responsibility (Loscocco, 1998)	-.21
Control Over Work Schedule (Iverson, Olekalns, & Erwin, 1998; Thomas & Ganster, 1995)	.35

All correlations greater than .17 in absolute value are statistically significant at $p < .05$.

RELIABILITY

Internal Consistency

Scale	α =
Informal Work Accommodations to Family	.79

Comments

- The IWAF scale proved to be reasonably valid and reliable in two separate samples.

- Some problems of the scale may be associated with the summation of items across broad behavioral constructs. The approach may have reduced inter-item correlations, introduced unsystematic variance, and served to attenuate relationships between the IWAF scale and hypothesized predictors. However, despite these issues, the IWAF scale was found to be valid and reliable.

- The relatively small sample sizes precluded factor analysis of the

Work Family/Work-Life Measures

TITLE OF MEASURE: INFORMAL WORK ACCOMMODATIONS TO FAMILY (IWAF)

IWAF scale items. Identification of an underlying factor structure could allow examination of linkages between particular IWAF factors and other relevant constructs.

- Sample 1 was drawn from a single organization which may have unique characteristics. Additional research is necessary to determine the external validity of the study findings across different populations and settings. For example, workers in blue-collar or high-customer-contact occupations may not have the option to use IWAF behaviors or may use them very differently.

Bibliography (studies that have used the measure)

Contact Information

Scott J. Behson, Chair
Department of Management
Samuel J. Silberman College of Business Administration
Fairleigh Dickinson University
1000 River Road (H-DH2-06)
Teaneck, New Jersey, NJ 07666, USA

Tel: 201-692-7233

e-mail: Behson@fdu.edu

www.scottbehson.homestead.com

Work Family/Work-Life Measures

TITLE OF MEASURE	JOB-FAMILY ROLE STRAIN SCALE
Source/Primary reference	Bohen, H., & Viveros-Long, A. (1981). *Balancing job and family life: Do flexible work schedules help?* Philadelphia: Temple University Press.
Construct measured	Perceptions of stress related to internalized values and emotions (worry, guilt, pressure, contentment, fulfillment, balance) in regard to job and family obligations
Brief description	The instrument includes 19 questions rated on a 5-point scale ranging from 1 = always to 5 = never. Items covered 5 dimensions (based on the work of Komarovsky, 1977): 1. Ambiguity about norms (3 items) 2. Socially structured insufficiency of resources for role fulfillment (3 items) 3. Low rewards for role conformity (3 items) 4. Conflict about normative phenomena (4 items) 5. Overload of role obligations (6 items) The scale is divided into two parts. The "Adult" part can be answered by participants with or without children. In the second part, "Parent," the items are relevant only for people with children.
Sample items	- I worry that other people at work think my family interferes with my job. - I worry about how my kids are when I am working. - I feel more respected than I would if I didn't have a job. - My work keeps me away from my family too much. - I feel I have more to do than I can handle comfortably.
Appropriate for whom (i.e. which population/s)	Employed persons with or without children
Translations & cultural adaptations available	None known

Work Family/Work-Life Measures

TITLE OF MEASURE JOB-FAMILY ROLE STRAIN SCALE

How developed

The development strategy for the Family Management Scale and the Role Strain Scale were coordinated. To develop the items for the scale, the authors followed three methods. First, they reviewed statements of family members who participated in five studies. Their statements were coded for areas of strain in performing family and work roles.

As a second strategy, individual and group conversations were held with 10 families. Parents were asked to report on the strains they experience when trying to be both good workers and good parents. Children described the types of strains faced by their parents.

In a third step, the statements developed from the prior two strategies were shown in written form to two groups of federal employees in two different agencies. They discussed whether each statement reflected their feelings and experience.

Psychometric properties

<u>STUDY SAMPLE</u>

Participants	Demographics	
Sample Size	$N = 706$	
Description	Women and men employed by one of two agencies of the U.S. federal government, working either standard time or "flexitime"	
Gender†	Standard Time	Flexitime
Female	49%	45%
Male	51%	55%
Race/Ethnicity	Standard Time	Flexitime
White	67%	70%
Minorities	33%	30%

†Gender and race/ethnicity are reported for the target sample, rather than the final study sample.

<u>VALIDITY</u>
Content Validity

To establish the initial content validity, six judges reviewed the items. They rated them according to the degree the items tapped the content designated for the scale. Items that were approved in this process were included in the scale.

Construct Validity

To establish the construct validity of the scale, a factor analysis was performed, using a principal components analysis with varimax rotation.

Work Family/Work-Life Measures

TITLE OF MEASURE: JOB-FAMILY ROLE STRAIN SCALE

The scale was considered in three parts, one for all adults, one for parents only, and a total scale combining the first two. The analyses showed that the items did not factor perfectly, but the three versions do have factorial clusters that coincide with five of the six of Komarovsky's (1977) modes that served as a theoretical basis for the scale.

Concurrent Validity

Respondents' scores on the scale were correlated with their score on a set of criterion variables. Positive correlations were found between the degree of role strain and the time spent working and commuting, the time spent at the job and in family work, as well as the perception of family-work interference.

	Adult Scale		Total Scale	
Criterion Variable	*n =*	*r =*	*n =*	*r =*
# of hours worked	567	.08*	273	.07
# of hours working and commuting	550	.16***	268	.11*
# of hours at job and in family work	243	.24***	242	.18**
Spouse works	838	.09*	221	.17**
Perception of family work interference	549	.49***	271	.52***
Age of youngest child	540	-.05	271	-.09
Family life-cycle stage	574	.10**	273	-.09
Outside help	430	-.06	267	-.03

$*p < .05; **p < .01; ***p < .001$

<u>RELIABILITY</u>

Internal Consistency

Pretest (n = 50):

Scale	α =
Job-Family Role Strain	.71

Posttest:

	Adult Scale		Total Scale	
Sample	*n =*	α =	*n =*	α =
Female	170	.67	66	.55
Male	263	.64	113	.53
Total	481	.72	212	.60

Work Family/Work-Life Measures

TITLE OF MEASURE   JOB-FAMILY ROLE STRAIN SCALE

Comments
- The measure is not directly about workplace issues. However, it is relevant to the broader topic of work-life integration, and as such can provide useful information.
- Assesses multiple aspects of role strain including ambiguity re: organizational norms, fit between personal values and role expectations, and role overload.

Bibliography (studies that have used the measure)

Contact Information

Halcyone H. Bohen
5357 Macarthur Blvd.
Washington, DC 20016-2539, USA

Tel: 202-364-0962

e-mail: halcybohen@aol.com

Work Family/Work-Life Measures

TITLE OF MEASURE	FAMILY MANAGEMENT SCALE
Source/Primary reference	Bohen, H., & Viveros-Long, A. (1981). *Balancing job and family life: Do flexible work schedules help?* Philadelphia: Temple University Press.
Construct measured	Feelings about the logistics of family life
Brief description	The scale is concerned with the routine and special activities that employed persons must manage outside their hours of work. The scale includes a list of 21 activities that are rated on a 5-point scale based on how difficult the respondent feels it is to manage each type of family responsibility. A higher score indicates more difficulties. It has some questions for all adults and some for parents only. It includes items regarding: <ul><li>health</li><li>education/child care</li><li>retail services</li><li>commuting</li><li>family interaction</li><li>community interaction</li><li>general overlapping items</li></ul>
Sample items	How difficult is it: <ul><li>To go to health care appointments</li><li>To go to school events for your children</li><li>To go shopping</li><li>To avoid rush hour</li><li>To visit or help neighbors or friends</li><li>To adjust your work hours to the needs of the other family members.</li><li>To go to work later that usual if you need to</li></ul>
Appropriate for whom (i.e. which population/s)	Employed persons with or without children
Translations & cultural adaptations available	None known

Work Family/Work-Life Measures

TITLE OF MEASURE	FAMILY MANAGEMENT SCALE

How developed

The development strategy for the Family Management Scale and the Role Strain Scale were coordinated. To develop the items for the scale, the authors followed three methods. First, they reviewed statements of family members who participated in five studies. Their statements were coded for areas of strain in performing family and work roles.

As a second strategy, individual and group conversations were held with 10 families. Parents were asked to report on the strains they experience when trying to be both good workers and good parents. Children described the types of strains faced by their parents.

In a third step, the statements developed from the prior two strategies were shown in written form to two groups of federal employees in two different agencies. They discussed whether each statement reflected their feelings and experience.

Psychometric properties

<u>STUDY SAMPLE</u>

Participants		*Demographics*	
Sample Size		$N = 706$	
Description		Women and men employed by one of two agencies of the U.S. federal government, working either standard time or "flexitime"	
Gender†		Standard Time	Flexitime
	Female	49%	45%
	Male	51%	55%
Race/Ethnicity†		Standard Time	Flexitime
	White	67%	70%
	Minorities	33%	30%

†Gender and race/ethnicity are reported for the target sample, rather than the final study sample.

<u>VALIDITY</u>

Content Validity

To establish the initial content validity, six judges reviewed the items. They rated them according to the degree the items tapped the content designated for the scale. Items that were approved in this process were included in the scale.

TITLE OF MEASURE: FAMILY MANAGEMENT SCALE

Construct Validity

To establish the construct validity of the scale, a factor analysis was performed, using a principal components analysis with varimax rotation. The scale was considered in three parts, one for all adults, one for parents only, and a total scale combining the first two. As anticipated, the items for the adult scale factored into 3 clusters and the items for the parent scale factored into 1 cluster. Items for the total scale did not cluster into 4 factors as expected, rather 5 factors emerged. However, the items still clustered in generally expected categories, with child care activities accounting for the greatest variance.

Respondents' scores on the scale were correlated with their scores on a set of criterion variables. Positive correlations were found between the family management scale and the hours worked, the time spent working and commuting, perception of family-work interference, as well as the number of children under 18 years living at home.

	Adult Scale		Total Scale	
Criterion Variable	$N=$	$r=$	$N=$	$r=$
# of hours worked	544	.18***	222	.18**
# of hours working and commuting	527	.24***	219	.23***
# of hours at job and in family work	228	.01	195	-.08
Spouse works	352	.02	173	-.09
Perception of family-work interference	528	.41***	220	.42***
# of children under 18 years living at home	542	.15***	219	.28***
Family life-cycle stage	553	.10*	222	.14*
Outside help	397	-.01	217	.83

*$p < .05$; **$p < .01$; ***$p < .001$

RELIABILITY

Internal Consistency

Pretest (N = 50):

Scale	$\alpha =$
Family Management	.93

Posttest:

	Adult Scale		Total Scale	
Sample	$N=$	$\alpha =$	$N=$	$\alpha =$
Female	208	.88	40	.92
Male	239	.89	52	.91
Total	449	.89	92	.92

Work Family/Work-Life Measures

TITLE OF MEASURE | FAMILY MANAGEMENT SCALE

Test-retest Reliability

Scale	Reliability Estimate
Family Management	.93

Comments

Bibliography (studies that have used the measure)

Contact Information

Halcyone H. Bohen
5357 Macarthur Blvd.
Washington, DC 20016-2539, USA

Tel: 202-364-0962

e-mail: halcybohen@aol.com

Work Family/Work-Life Measures

TITLE OF MEASURE	SPILLOVER BETWEEN HOME AND JOB RESPONSIBILITIES
Source/Primary reference	Cedillo-Becerril, L. (1999). *Psychosocial risk factors among women workers in the maquiladora industry in Mexico*. Doctoral dissertation, Dept. of Work Environment, University of Massachusetts Lowell.
Construct measured	Lack of balance between job and family responsibilities
Brief description	The approach is a 2-item measure of work-family and family-work interference in Spanish, developed for research in Mexico.
Sample items	▪ El tiempo que dedica a su trabajo asalariado ¿le impide cumplir totalmente con sus obligaciones domésticas? (Time required by your job duties does not allow you to accomplish home responsibilities.) ▪ El tiempo que necesita para cumplir totalmente con sus obligaciones domésticas ¿le impide cumplir totalmente con su trabajo asalariado? (Time required by your home duties does not allow you to accomplish job responsibilities.)
Appropriate for whom (i.e. which population/s)	Working populations
Translations & cultural adaptations available	Spanish (original); available in English Mexican culture (original); not tested in another culture
How developed	The two general questions were written by the researcher after individual and group interviews with women workers pointed out some worries about balancing job and family responsibilities.
Psychometric properties	<u>STUDY SAMPLE</u> Questions were designed and applied as a part of a questionnaire answered by 370 Mexican women workers. | Participants	Demographics
Sample Size	$n = 370$
Description	Mexican women workers	

Work Family/Work-Life Measures

TITLE OF MEASURE | SPILLOVER BETWEEN HOME AND JOB RESPONSIBILITIES

VALIDITY

Construct Validity

Factor analyses were conducted on 16 items, resulting in only 1 factor with two items: loading of .84; communality of 0.72. The authors actually developed 5 different scales, but only the one shown to have the highest predictive validity is reported here. However, two of the other scales related to conflicting relationships had somewhat high predictive validity as well.

RELIABILITY

Internal Consistency

Scale	Cronbach's α =
Spillover Between Home and Job Responsibilities	.67

Comments

- The scale was associated with three psychological strain indicators: OR = 1.56, 1.58, and 2.33 for depression, anger and exhaustion (controlling for 4 non-work stressors in multivariable logistic regression models).

- The scale showed good psychometric and predictive properties. However, the authors opined that additional items should be developed to strengthen it, since it was originally intended to include 5 items.

Bibliography (studies that have used the measure)

There are two ongoing studies that are using the scale reported here:

Scarone, M. *Trabajo y tensión psicológica: factores psicosociales de riesgo para la salud de las trabajadoras del servicio telefónico. Estudio de la interacción cliente trabajadora.* Tesis de Maestria en Ciencias Sociales, area de Relaciones Industriales. El Colegio de Sonora, Mexico.

Torres A. L. *Evaluación macro-ergonómica y estrés durante el embarazo en mujeres derechohabientes del Instituto Mexicano del Seguro Social.*

Contact Information

Leonor Cedillo
e-mail: leonor_cedillo@yahoo.com
No cost

Work Family/Work-Life Measures

TITLE OF MEASURE	WORK-FAMILY POLICIES-PERCEIVED MANAGEMENT SUPPORT AND USABILITY
Source/Primary reference	Eaton, S. C. (1999). Gender and the structure of work in biotechnology. *The Annals of the New York Academy of Science, 865*, 175-188.
Construct measured	The extent to which the organization supports employee efforts to balance work and family
Brief description	One section includes 7 items about perceptions of organizational policies and expectations about work-life issues. Each item is rated on a 5-point scale from "not at all" to "a great deal." A separate scale covers the formal and informal availability and the usability of 10 company policies and programs that assist in balancing work and family.
Sample items	Section 1: - Do your managers have a good understanding of people's work and family needs? - Does your company expect employees to keep family matters out of the workplace? - Are you expected to work long hours on short notice? - Do you need to negotiate individually with your supervisor when you have a personal life concern that might conflict with your work? - Do you worry that requesting time off for personal reasons will hurt your career? Section 2: Scales of Formal W/F Practices, Informal W/F Practices, and Perceived Usability (alternately called "PERC" and "USABLE") For each flexibility policy or benefit listed below, please indicate: - Whether it is formally available? - Whether it is informally available? - Whether, if it is available, you feel free to use it? Ten policies are then listed, including flextime, job sharing, and use of sick days to care for children.
Appropriate for whom (i.e. which population/s)	Adult workers

Work Family/Work-Life Measures

TITLE OF MEASURE	WORK-FAMILY POLICIES-PERCEIVED MANAGEMENT SUPPORT AND USABILITY

Translations & cultural adaptations available	None known
How developed	Based on interviews with human resource personnel and others, the primary author identified seven practices that could potentially affect work-life balance that are related to organizational flexibility. These became the basis for the survey items in Section 1.
Psychometric properties	*STUDY SAMPLE* Interviews and surveys were conducted with employees (*n* = 461) in seven biopharmaceutical firms in one state, ranging from quite large (over 1,000 employees) to small (fewer than 100). Most participants were well-educated (college or graduate degree) and held professional or managerial positions with mean household incomes around $70,000.

	Women *n* = 253 (56%)		Men *n* = 200 (44%)	
	Median	*Range*	*Median*	*Range*
Age	35.6	22-59	37.7	19-68
Years of Service	4.6	0-14	4.8	0-16
	n	(%)	*n*	(%)
Employed Full-Time	235	94%	197	100%
Married or partnered	176	64%	151	78%
One or more children	110	40%	109	56%

The ethnic/racial make-up of the sample was not reported.

VALIDITY

Content Validity

Open-ended interview questions and data from in-depth company case studies (observations, group discussions, focused interviews) (Eaton & Bailyn, 1999) were used to explore these issues in depth and to generate survey questions.

The means and standard deviations were higher for informal than for formal policies. The author interpreted these findings as evidence of face validity, in that flexibility is likely to be available either through formal company policies or through informal work group arrangements and the variability in such arrangements within a company or industry should therefore be higher.

Construct Validity

Scores on the USABLE index were higher for managers, which would be consistent with a higher degree of job control.

Comments	<ul><li>Goes beyond theoretical availability to address a specific feature of the work environment, namely whether policies are really experienced as accessible to employees</li><li>In cross-sectional data, the USABLE index was positively associated with organizational commitment; consistent with expectations, this association was weaker among employees who experienced a higher degree of control over their work pace and scheduling.</li><li>More psychometric assessment is needed in general.</li><li>In particular, since the ethnic/racial make-up of the sample was not reported, it would be useful to assess the scale's validity and reliability for multiple ethnic/racial groups.</li></ul>
Bibliography (studies that have used the measure)	Eaton, S. C. (1998). Gender and work in biotech firms. *Radcliffe Quarterly, 84*(2), 25. Eaton, S. C. (1999). Work and family practices in biotech firms. In P. Voos, (Ed). *Proceedings of the Industrial Relations Research Association Annual Meeting*, New York, 1, 8-14. Eaton, S. (2003). If you can use them: Flexibility, policies, organizational commitment and perceived performance. *Industrial Relations*, 145-167. Eaton, S. C. & Bailyn, L. (1999). Emergent career paths in changing organizations: Work and life strategies of professionals in biotechnology firms. *Annals of the American Academy of Political and Social Science, 562*, 159-173.
Contact Information	Susan Eaton e-mail: Susan_Eaton@ksg.harvard.edu

Work Family/Work-Life Measures

TITLE OF MEASURE	EMPLOYER SUPPORT FOR FAMILY
Source/Primary reference	Friedman, S. & Greenhaus, J. (2000). *Work and family: Allies or enemies*. New York, NY. Oxford University Press.
Construct measured	Organizational support for work and family balance
Brief description	The instrument includes 5 items about respondents' perceptions of the support employees in general receive for balancing work and family responsibilities. Each item is rated on a 5-point scale from strongly disagree to strongly agree.
Sample items	■ The level of commitment expected by my organization requires that employees choose between advancing their careers and devoting time to their families. (reverse score) ■ My organization is understanding when employees have a hard time juggling work and family responsibilities. ■ Career advancement is jeopardized if employees do not accept assignments because of their family responsibilities. (reverse score) ■ My organization has a satisfactory family leave policy. ■ My organization allows for flexibility in work scheduling.
Appropriate for whom (i.e. which population/s)	Employed persons, with or without children
Translations & cultural adaptations available	None known
How developed	Items were developed by authors based on a review of relevant literature.
Psychometric properties	*STUDY SAMPLE*

Participants	Demographics	
Sample Size	n = 861	
Description	Employed alumni from two business schools	
Age	Mean	38.4
Gender	Female	33.8%
	Male	66.2%
Race/Ethnicity	Caucasian	92.6%
Marital Status	Married	75.6%
	Not Married	24.4%
Have Children		57.5%

Work Family/Work-Life Measures

TITLE OF MEASURE: EMPLOYER SUPPORT FOR FAMILY

RELIABILITY

Internal Consistency

Scale	α =
Employer Support for Family	.78

Comments

- Looks at perceptions of organizational support and values (i.e., adds an assessment of the organizational context that complements many of the other work-life measures that tap individuals' overload, stress, and/or role conflict).

- The scale was developed with a primarily white sample. It would be useful to assess its validity and reliability for multiple ethnic/racial groups.

Bibliography (studies that have used the measure)

Contact Information

Jeffrey H. Greenhaus
William A. Mackie Professor
Department of Management
Drexel University
Philadelphia, PA 19104, USA

e-mail: jhg23@drexel.edu

Stewart Friedman

e-mail: Friedman@wharton.upenn.edu

Work Family/Work-Life Measures

TITLE OF MEASURE	WORK-FAMILY INTERFERENCE AND TRADEOFFS
Source/Primary reference	Friedman, S. & Greenhaus, J. (2000). *Work and family: Allies or enemies.* New York, NY. Oxford University Press.
Construct measured	The perception that the demands of the work role and the demands of the family role interfere with one another.
Brief description	The instrument includes 11 items organized into 3 subscales (all rated on a 5-point scale from strongly disagree to strongly agree): 1. Behavioral Work Interference with Family – 2 items 2. Work Interference with Family - 4 items 3. Family Interference with Work - 5 items Two additional items ask respondents about pressures to decide between career and family (the first of which was reverse scored).
Sample items	Behavioral Work Interference with Family - My partner complains that I treat family members as if they are work associates or subordinates. - I find it difficult making the transition from my job to home life. Work Interference with Family - When I spend time with my family I am bothered by all the things on the job that I should be doing. - Because of my job responsibilities, the time I spend with my family is less enjoyable and more pressured. - Because of my job responsibilities I have to miss out on home or family activities in which I should participate. - Pursuing a demanding career makes it difficult for me to be an attentive spouse/partner. Family Interference with Work - When I spend time on my job, I am bothered by all the things I should be doing with my family. - The demands of family life interfere with achieving success in my career. - Being a parent limits my career success.

Work Family/Work-Life Measures

TITLE OF MEASURE — WORK-FAMILY INTERFERENCE AND TRADEOFFS

- Because of my family responsibilities, I have to turn down job activities or opportunities that I should take on.
- Because of my family responsibilities, the time that I spend on my job is less enjoyable and more pressured.

Tradeoffs
- I can "have it all" (a rewarding career, satisfying family relationships and a fulfilling personal life).
- The conflicting demands of career and family require that I decide which is more important.

Appropriate for whom (i.e. which population/s)
Employed persons, with or without children

Translations & cultural adaptations available
None known

How developed
Items were developed by authors based on a review of relevant literature.

Psychometric properties

STUDY SAMPLE

Participants	Demographics	
Sample Size	n = 861	
Description	Employed alumni from two business schools	
Age	Mean	38.4
Gender	Female	33.8%
	Male	66.2%
Race/Ethnicity	Caucasian	92.6%
Marital Status	Married	75.6%
	Not Married	24.4%
Have Children		57.5%

RELIABILITY

Internal Consistency

Subscale	α =
Behavioral Work Interference with Family	.68
Work Interference with Family	.73
Family Interference with Work	.78
Tradeoffs	.58

Work Family/Work-Life Measures

TITLE OF MEASURE	WORK-FAMILY INTERFERENCE AND TRADEOFFS
Comments	The scale was developed with a primarily white sample. It would be useful to assess its validity and reliability for multiple ethnic/racial groups.
Bibliography (studies that have used the measure)	
Contact Information	Jeffrey H. Greenhaus William A. Mackie Professor Department of Management Drexel University Philadelphia, PA 19104, USA e-mail: jhg23@drexel.edu Stewart Friedman e-mail: Friedman@wharton.upenn.edu

Work Family/Work-Life Measures

TITLE OF MEASURE	ADJUSTMENT OF WORK SCHEDULE
Source/Primary reference	Friedman, S. & Greenhaus, J. (2000). *Work and family: Allies or enemies.* New York, NY. Oxford University Press.
Construct measured	Adjustment of work schedule for family and personal reasons
Brief description	The instrument includes 4 items about respondents' perceptions of the frequency with which the respondent has adjusted or limited his or her work schedule to meet family or personal needs over the last two years. Each item is rated on a 5-point scale from never to frequently.
Sample items	Within the last two years, how often have you: • Adjusted your hours of arrival and departure from work to suit your personal and family activities. • Structured your hours at work in order to be home at certain specific times. • Limited the time you devoted to work during weekends. • Limited the time you devoted to work-related travel.
Appropriate for whom (i.e. which population/s)	Employed persons, with or without children
Translations & cultural adaptations available	None known
How developed	Items were developed by authors based on a review of relevant literature.
Psychometric properties	*STUDY SAMPLE*

Participants	Demographics	
Sample Size	n = 861	
Description	Employed alumni from two business schools	
Age	Mean	38.4
Gender	Female	33.8%
	Male	66.2%
Race/Ethnicity	Caucasian	92.6%
Marital Status	Married	75.6%
	Not Married	24.4%
Have Children		57.5%

Work Family/Work-Life Measures

TITLE OF MEASURE | ADJUSTMENT OF WORK SCHEDULE

RELIABILITY

Internal Consistency

Scale	α =
Adjustment of Work Schedule	.70

Comments
- The scale was developed with a primarily white sample. It would be useful to assess its validity and reliability for multiple ethnic/racial groups.

Bibliography (studies that have used the measure)

Contact Information

Jeffrey H. Greenhaus
William A. Mackie Professor
Department of Management
Drexel University
Philadelphia, PA 19104, USA

e-mail: jhg23@drexel.edu

Stewart Friedman

e-mail: Friedman@wharton.upenn.edu

Work Family/Work-Life Measures

TITLE OF MEASURE	**WORK-FAMILY CONFLICT**
Source/Primary reference	Frone M., & Yardley, J. K. (1996). Workplace family-supportive programs: Predictors of employed parents' importance ratings. *Journal of Occupational and Organizational Psychology, 69*(4), 351-367.
Construct measured	Interference of the employed adults' family roles with their work roles and responsibilities.
Brief description	Twelve items were used to assess work-family conflict; six items each assessed the degree to which a respondent's job interferes with his or her home life (work-[>]family conflict) and the degree to which a respondent's home life interferes with his or her job (family-[>]work conflict). A 5-point response scale was used with 1 = never to 5 = very often.
Sample items	**Work-[>]family conflict** - After work, I come home too tired to do some of the things I'd like to do. - On the job I have so much work to do that it takes away from my personal interests. - My family/friends dislike how often I am preoccupied with my work while I am at home. - My work takes up time that I'd like to spend with family/friends. **Family-[>]work conflict** - I'm too tired at work because of the things I have to do at home. - My personal demands are so great that it takes away from my work. - My superiors and peers dislike how often I am preoccupied with my personal life while at work. - My personal life takes up time that I'd like to spend at work.
Appropriate for whom (i.e. which population/s)	Working adults
Translations & cultural adaptations available	None known

Work Family/Work-Life Measures

TITLE OF MEASURE — **WORK-FAMILY CONFLICT**

How developed

Prior research suggests that family demands affect job outcomes indirectly when family demands spill over into work time/tasks, whereas work demands affect family outcomes when work demands conflict with family (see Frone, Russell, & Cooper, 1992a, model of the work-family interface).

For each of the two dimensions of work-family conflict, the present measure was developed by combining the two-item scale developed by Frone, Russell, & Cooper (1992a,b) and the four-item scale used by Gutek, Searle & Klepa (1991).

Psychometric properties

STUDY SAMPLE

The sample was drawn from a mid-sized financial services company located in Ontario, Canada. A questionnaire covering a variety of issues regarding work and family life was distributed to all 600 employees. The subsample for the present study was composed of the 252 respondents who had children living at home.

Participants	Demographics	
Sample Size	n = 252	
Description	Employees of a mid-sized financial services company located in Ontario, Canada with children living at home	
Age	Mean (SD)	36.17 (6.19)
Gender	Female	74%
	Male	26%
Race/Ethnicity		Not reported
Education	College	45.7%
	High School	53.2%
	Less than High School	1.2%
Income	Median Family (Canadian)	$50,000-$59,999
Years with Company	Mean (SD)	8.98 (6.60)
Marital Status	Married/Living as Married	90.5%
Number of Children Living at Home	Mode	2.0
	Range	1 - 5

A questionnaire covering a variety of issues regarding work and family life was filled out on company time. Respondents were informed that the primary purpose of the questionnaire was for an outside research project examining job stress and work-family processes. A secondary goal was to provide feedback to the company regarding the work-family problems and needs of its employees.

Work Family/Work-Life Measures

TITLE OF MEASURE — WORK-FAMILY CONFLICT

VALIDITY

Construct Validity

To assess the dimensionality of the work-family conflict items, an exploratory factor analysis was conducted. The factor analysis revealed three factors with eigen values greater than or equal to 1.0. However, the scree plot suggested retaining only two factors. A two-factor solution revealed that the six work-[>] family conflict items loaded highly on the first factor (oblique rotated loadings = .47 to .90), whereas the six family-[>] work items loaded highly on the second factor (oblique rotated loadings = .46 to .74). In addition, the 12 cross-factor loadings were small (oblique rotated loadings = -.06 to. 19).

RELIABILITY

Internal Consistency

Scale	$\alpha =$
Work-[>] family conflict	.87
family-[>] work conflict	.79

Comments

The ethnic/racial make-up of the sample was not reported. It would be useful to assess the scale's validity and reliability for multiple ethnic/racial groups.

Bibliography (studies that have used the measure)

Contact Information

Michael R. Frone
Research Institute on Addictions
University at Buffalo
1021 Main Street
Buffalo, NY 14203-1016, USA

Tel: 716-887-2566

e-mail: frone@ria.buffalo.edu

Work Family/Work-Life Measures

TITLE OF MEASURE	SURVEY WORK-HOME INTERACTION-NIJMEGEN (SWING)
Source/Primary reference	Geurts, S., Taris, T., Kompier, et al. (in preparation). SWING: Development and validation of the 'Survey Work-home Interaction-Nijmegen' in five different occupational groups. Available from Sabine Geurts at S.Geurts@psych.kun.nl.
Construct measured	The extent to which one's functioning in one domain is influenced by demands from the other domain.
Brief description	The instrument consists of 27 items, measured on 4-point scales from 0 = (almost) never to 3 = (almost) always. This instrument taps four types of work-home interaction (WHI): 1. Work negatively influencing home (WHI-) 2. Home negatively influencing work (HWI-) 3. Work positively influencing home (WHI+) 4. Home positively influencing work (HWI+)
Sample items	How often does it happen that . . . • You are irritable at home because your work is demanding? • The situation at home makes you so irritable that you take your frustrations out on your colleagues? • You come home cheerfully after a successful day at work, positively affecting the atmosphere at home? • After spending time with your spouse/family/friends, you go to work in a good mood, positively affecting the atmosphere at work?
Appropriate for whom (i.e. which population/s)	Employed adults
Translations & cultural adaptations available	English and Dutch versions are available. There is an additional shortened version in German.
How developed	The authors reviewed 21 existing scales that focus on work-home interactions. From a pool of 214, items were chosen that met the following criteria: • Fit the definition of WHI (having a clear direction with the cause in one domain and effect in the other domain). • Are not confounded with outcome measures.

TITLE OF MEASURE	*Work Family/Work-Life Measures* SURVEY WORK-HOME INTERACTION-NIJMEGEN (SWING)

- Are not confounded with demands from work or home domains.

A team of researchers chose the items appropriate to be included in SWING, and when the number of items was too small to cover a dimension, new items were developed.

Psychometric properties

STUDY SAMPLES

Participants	Sample 1	Sample 2	Sample 3
Sample Size	n = 751	n = 524	n = 624
Description	Employees of the Dutch Postal Services	Employees from a manufacturing company in the electronic industrial sector	Employees from a financial consultancy firm
Gender	Not available	Not available	Not available
Race/Ethnicity	Not available	Not available	Not available

VALIDITY

Construct Validity

To examine the construct validity of the Dutch SWING, the four subscales were related to relevant work and home characteristics. In Sample 1 workload and job control were measured with the two scales from the extensively validated Dutch Questionnaire of Experience and Evaluation of Work (Van Veldhoven, Meijman, Broersen, & Fortuin, 1997).

In Samples 2 and 3 the measures of workload and job control were based on the well-known Job Content Questionnaire (JCQ) of Karasek (1985). The measure of job support was measured by four items derived from the Questionnaire of Organizational Stress-Doetinchem (VOS-D; Bergers, Marcelissen, & Wolff, 1986). The home variables were for the largest part self-developed. The measure of home support was derived from Peeters (1994).

	Sample 1				
Interaction Type	*Work load*	*Job control*	*Job support*	*Household tasks*	*Home support*
WHI-	.56	-.27	-.32	-	-
HWI-	.17	-	-.15	.12	-
WHI+	-	.11	.16	-	-
HWI+	-	-	.16	-	-

Note: '-' refers to non-significant correlations or correlations < = .10

Work Family/Work-Life Measures

TITLE OF MEASURE： SURVEY WORK-HOME INTERACTION-NIJMEGEN (SWING)

	\multicolumn{6}{c}{Sample 2}
Interaction Type	Work load	Job control	Job support	Workload at home	Household tasks	Home support
WHI-	.40	-.13	-.27	.28	-	-
HWI-	-	-.23	-.16	.34	.16	-.16
WHI+	-	-	-	-	.11	-
HWI+	-.11	-	.11	.13	.13	-

Note: '-' refers to non-significant correlations or correlations < .10

	\multicolumn{4}{c}{Sample 3}
Interaction Type	Work load	Job control	Job support	Household tasks
WHI-	.47	-	-.25	-.27
HWI-	-	-.13	-	-
WHI+	-	-	.12	-
HWI+	-	-	-	.14

Note: '-' refers to non-significant correlations or correlations < .10

Factor analysis shows that the four subscales are fairly independent of one another.

	Sample 1			Sample 2			Sample 3		
Type	HWI-	WHI+	HWI+	HWI-	WHI+	HWI+	HWI-	WHI+	HWI+
WHI-	.35	.23	.11	.34	.02	-.02	.27	.08	-.03
HWI-	-	.12	.08	-	.14	.11	-	.11	.11
WHI+		-	.43		-	.55		-	.62
HWI+			-			-			-

RELIABILITY

Internal Consistency

	Sample 1			Sample 2			Sample 3		
Interaction Type	M	SD	α	M	SD	α	M	SD	α
WHI-	.81	.50	.88	.96	.46	.84	.92	.45	.86
HWI-	.38	.34	.77	.50	.42	.82	.50	.34	.73
WHI+	.76	.50	.75	.99	.52	.72	.98	.53	.80
HWI+	1.20	.78	.82	1.16	.67	.84	1.29	.62	.78

Work Family/Work-Life Measures

TITLE OF MEASURE: SURVEY WORK-HOME INTERACTION-NIJMEGEN (SWING)

Comments	Documentation of relationships to health

Fatigue was measured with a subscale from the Checklist Individual Strength (CIS; Vercoulen, Alberts, & Bleijenberg, 1999). Health complaints were measured by the VOEG (13-item version; Joosten & Drop, 1987), but with exclusion of four items that referred to fatigue. The sleep quality measure was based on the Groninger Sleep Quality Scale (GSKS, Meijman et al., 1988). Depressive mood was measured by a short version of the CES-D (Kohout et al., 1993; Radloff, 1977).

	Sample 1			*Sample 2*			*Sample 3*
Type	*Fatigue*	*Health complaints*	*Sleep quality*	*Fatigue*	*Depressive mood*	*Sleep quality*	*Fatigue*
WHI-	.52	.38	-.38	.49	.52	-.46	.47
HWI-	.24	.17	-.24	.44	.40	-.29	.29
WHI+	-	-	-	-	-	-	-
HWI+	-	-	-	-	-	-	-

Note: '-' refers to non-significant correlations or correlations < .10

- There are several strengths of this instrument:
    - one of the few instruments that measure not only negative but also positive interaction between work and home
    - based on a theoretical framework
    - applicable to all employed workers (those with and without partner or children)
- Some disadvantages include:
    - Quite long (27 items), although short version (16 items) is available, and the four subscales can be used apart from one another.
    - Mean scores on the four subscales are rather low.
    - Relationship with demands in home situation is not completely clear.

Bibliography (studies that have used the measure)	Demerouti, E., Bakker, A. B., & Bulters, A. J. (in press, 2003). The loss spiral of work pressure, work-home interference and exhaustion: Reciprocal relations in a three-wave study. *Journal of Vocational Behavior*.

Work Family/Work-Life Measures

TITLE OF MEASURE — **SURVEY WORK-HOME INTERACTION-NIJMEGEN (SWING)**

Dikkers, J., Den Dulk, L., Geurts, S., & Peper, B. (in press). Work-life arrangements and fatigue in two Dutch organizations. In S. Poelmans (Ed.), *Work and family: An international research perspective*, Lawrence Erlbaum.

Geurts, S, & Demerouti, E. (2003). Work/non-work interface. A review of theories and findings. In M. J. Schabracq, J. A. M. Winnubst, & C. L. Cooper (Eds.), *The Handbook of Work & Health Psychology* (pp. 279-312). Chichester: John Wiley & Sons.

Montgomery, A., Peeters, M. C. W., & Schaufeli, W. B. (2003). Work-home interference among newspaper managers: Its relationship with burnout and engagement. *Anxiety, Stress & Coping 16*(17):195-211.

Van der Hulst, M. & Geurts, S. (2001) Associations between overtime and psychological health in high-and low-reward jobs. *Work & Stress, 15*, 227-240.

Contact Information

Dr. S.A. Geurts
University of Nijmegen
Department of Work & Organizational Psychology
P.O. Box 9104
6500 HE Nijmegen, Netherlands

e-mail: S.Geurts@psych.kun.nl

Work Family/Work-Life Measures

TITLE OF MEASURE	WORK-FAMILY CONFLICT
Source/Primary reference	Gutek, B. A., Searle, S. & Klepa, L. (1991). Rational versus gender role explanations for work-family conflict. *Journal of Applied Psychology*, 76(4), 560-568.
Construct measured	Extent to which work demands interfere with family and family demands interfere with work
Brief description	The instrument consists of 8 items—4 items were developed to measure work interference with family (WIF) and 4 times were developed to measure family interference with work (FIW). The response options for both sets of questions were 5-point scales ranging from 1 = strongly agree to 5 = strongly disagree.
Sample items	▪ After work, I come home too tired to do some of the things I'd like to do. (WIF) ▪ On the job I have so much work to do that it takes away from my personal interests. (WIF) ▪ I'm often too tired at work because of things I have to do at home. (FIW) ▪ My personal demands are so great that it takes away from my work. (FIW)
Appropriate for whom (i.e. which population/s)	Working adults
Translations & cultural adaptations available	None known
How developed	The scale was developed by combining items from two previously developed scales. Four items developed by Kopelman, Greenhaus, and Connoly (1983) assessed work-interference-with family (WIF). Another four items, paralleling the WIF items, were developed by Burley (1989) to assess family-interference-with-work (FIW). In Gutek's work, the items were reverse coded so that a high score would represent high conflict.
Psychometric properties	See Kopelman Scale entry for Kopelman's items. Psychometric information on Burley items is not available.

Work Family/Work-Life Measures

TITLE OF MEASURE   WORK-FAMILY CONFLICT

STUDY SAMPLES

Participants		Study 1	Study 2
Sample Size		n = 534†	n = 209‡
Description		Psychologists who were full members or fellows of at least one of APA Divisions 9 or 35	Senior Managers
Average Age	Women	47	39
	Men	50	46
Gender	Female	69.6%	25%
	Male	30.4%	75%
Race/Ethnicity	Women: White	Not reported	87%
	Men: White	Not reported	82%

†A subsample including only those participants who lived with at least one other family member (spouse, significant other of either sex, or a child) (n = 423) was used for all analyses. Among these respondents, 65% had at least one child living with them.

‡A subsample including only those participants who lived with at least one other family member (spouse, significant other of either sex, or a child) was used for all analyses. This subsample included 135 men and 43 women.

VALIDITY

Construct Validity

A factor analysis with varimax rotation revealed that the items for the two scales loaded on separate factors. The correlation between the two conflict scales was .26 showing that WIF and FIW are distinct.

RELIABILITY

Internal Consistency

Scale	Study 1 $\alpha =$	Study 2 $\alpha =$
WIF	.81	.83
FIW	.79	.83

Comments

- The scale represents an easy-to-use combination and refinement of scales developed by others.

Work Family/Work-Life Measures

TITLE OF MEASURE	WORK-FAMILY CONFLICT
Bibliography (studies that have used the measure)	Beutell, N. J., & Wittig-Berman, U. (1999). Predictors of work-family conflict and satisfaction with family, job career and life. *Psychological Reports, 85*(3), 893-904. Frone, M., Russell, M., & Cooper, M. (1996). Workplace family-supportive programs: Predictors of employed parents' importance ratings. *Journal of Occupational and Organizational Psychology, 69*, 351-366. Leiter, M. P.; & Durup, M. J. (1996). Work, home, and in-between: A longitudinal study of spillover. *Journal of Applied Behavioral Science, 32*(1), 29-48. Netemeyer, R. G.; & McMurrian, R. (1996). Development and validation of work-family conflict and family-work conflict scales. *Journal of Applied Psychology, 81*(4), 400-410. Parasuraman, S., Purohit, Y. S., Godshalk, V. M., & Beutell, N. J. (1996). Work and family variables, entrepreneurial career success, and psychological well-being. *Journal of Vocational Behavior, 48*, 275-300.
Contact Information	Barbara Gutek Department of Psychology University of Michigan Ann Arbor, MI 48109, USA

Work Family/Work-Life Measures

TITLE OF MEASURE	WORK-FAMILY INTERFERENCE
Source/Primary reference	Hughes, D. & Galinski, E. (1994). Gender, job and family conditions, and psychological symptoms. *Psychology of Women Quarterly, 18*(2), 251-271.
Construct measured	Work-family interference
Brief description	The measure consists of two subscales rated on a 5-point scale (from 1 = never to 5 = very often): 1. *Family role difficulty* subscale consists of 8 items that tap the family role difficulties that are attributed to the job. The items focus on issues such as time spent with family and difficulties with accomplishing logistical tasks. 2. *Job role difficulty* subscale consists of 6 items that address family role responsibilities that can contribute to difficulties at work. The items focus on frequency with which family responsibilities cause difficulties in accomplishing work roles. A global question requested respondents to give their perception of family and work interference.
Sample items	• Because of my job, it is difficult for me to spend enough time with my spouse. • Because of my family responsibilities, it is difficult for me to get to work on time. • All in all, how much would you say your work and family life interfere with each other?
Appropriate for whom (i.e. which population/s)	Employed adults
Translations & cultural adaptations available	No
How developed	Items were developed by the authors based on issues that emerged at work-family workshops in corporate settings as well as conceptual distinctions regarding the work-family interface in theoretical literature.
Psychometric properties	<u>STUDY SAMPLE</u> The participants were employees of a company in the Northeast. 90% of the respondents were white, 3% African American, 5% Asian, and 2%

| TITLE OF MEASURE | WORK-FAMILY INTERFERENCE |

Hispanic. The sample was divided into three categories: employed men with nonemployed spouses; employed men with employed spouses; and employed women with employed spouses.

Participants		Single-Earner Men	Dual-Earner Men	Dual-Earner Women
Sample Size		n = 142	n = 126	n = 161
Age	Range	21-67	21-67	22-64
	Mean	45	42	35
Race/Ethnicity	White	90%		
	African American	3%		
	Asian	5%		
	Hispanic	1%		
Have children under 17 years		61%	41%	43%

VALIDITY

Concurrent Validity

The correlation between the first subscale and the global question was high; the second subscale was only moderately correlated.

Subscale	Global question
Family Role Difficulty	r = .72
Job Role Difficulty	r = .42

RELIABILITY

Internal Consistency

Subscale	Cronbach's α =
Family Role Difficulty	.90
Job Role Difficulty	.83

Comments

The relationship to general health was not examined.

- The assessment is based on respondents' self-reports. Relies on respondents to make attributions about the causes of their role difficulties or psychological states.

- The study sample was predominantly white. It would be useful to assess the scale's validity and reliability for women and for multiple ethnic/racial groups.

Work Family/Work-Life Measures

TITLE OF MEASURE   WORK-FAMILY INTERFERENCE

Bibliography (studies that have used the measure)

Contact Information   Dianne Hughes
Department of Psychology
New York University
6 Washington Place
New York, NY 10003, USA

Work Family/Work-Life Measures

TITLE OF MEASURE	INTERROLE CONFLICT SCALE
Source/Primary reference	Kopelman, R. E., Greenhaus, J. H., & Connoly, T. F. (1983). A model of work, family, and interrole conflict: A construct validation study. *Organizational Behavior and Human Performance, 32*(2), 198-215.
Construct measured	Conflict between work and family roles
Brief description	The measure includes 8 items to assess the extent of conflict between work and family roles (i.e., perceptions of pressures within one role that are incompatible with pressures that arise within another role). Each item is rated on a 5-point scale that ranges from 1 = strongly disagree to 5 = strongly agree, where the higher the score, the higher the conflict.
Sample items	<ul><li>My work schedule often conflicts with my family life.</li><li>After work I come home too tired to do some of the things I'd like to do.</li><li>My family dislikes how often I am preoccupied with my work while I am home.</li><li>The demands of my job make it difficult to be relaxed all the time at home.</li></ul>
Appropriate for whom (i.e. which population/s)	Employed adults
Translations & cultural adaptations available	None known
How developed	The items for the interrole conflict scale were based on previous research that had identified seven types of work-family conflict, with three being the most prevalent: excessive work time, schedule conflicts, fatigue, and irritability. Based on these findings, 4 items were drafted, three addressing excessive work and one fatigue. In the second study, in addition to modifying wording of one item, four more items were added: 2 for excessive work demands, 1 for fatigue and irritability, and 1 for schedule conflicts.

Work Family/Work-Life Measures

TITLE OF MEASURE   INTERROLE CONFLICT SCALE

Psychometric properties

STUDY SAMPLES

Participants		Study 1	Study 2
Sample Size		n = 181	n = 91
Description		Alumni of a technical college	Students employed full-time
Age (M (SD))		43.3 (10.6)	35.6 (9.3)
Gender: Male		100%	50%
Race/Ethnicity		Not reported	Not reported
Education	Advanced Degree	51%	18%
	College Education	100%	56%
Organizational Tenure (M (SD))		11.5 (9.9) years	6.9 (6.4) years
Positional Tenure (M (SD))		4.6 (5) years	4.1 (4.5) years
Married		100%	99%
Spouse Employed		39%	82%
Have Children		84%	57%

VALIDITY

Construct Validity

Interrole conflict was one of three factors that emerged from analysis of a broader set of items designed to also assess work conflict and family conflict (i.e., incompatible pressures within the work and family domains).

	Median Factor Loading	
Scale	Study 1	Study 2
Interrole Conflict	.61	.65
Work Conflict	.46	.74
Family Conflict	.54	.57

Intercorrelations between scales (Pearson product-moment coefficients):

Scale	1	2	3
1 Interrole Conflict		0.30	0.30
2 Work Conflict	0.36		0.30
3 Family Conflict	0.22	0.14	

Above diagonal = Study 1

Below diagonal = Study 2

Work Family/Work-Life Measures

TITLE OF MEASURE: INTERROLE CONFLICT SCALE

RELIABILITY
Internal Consistence

Cronbach's alpha scores were high for each of the three scales in both studies.

	Study 1	Study 2
Interrole Conflict	α = 0.70	α = 0.89
Work Conflict	α = 0.70	α = 0.80
Family Conflict	α = 0.65	α = 0.87

Comments

- The validation studies had small samples, and in the second study, a convenience sample was used.

- This measure is probably the most frequently used in the formal research literature—sometimes as a full scale and sometimes by taking a subset of items and combining them with items from other sources (Allen, Herst, Bruck, & Sutton, 2000). See Gutek, Searle, & Klepa (1991) as an example.

- The scale was developed with a predominantly male sample. It is possible that factor loadings would have been different, and thus different items might have been retained, with a female sample.

- A researcher who recently used this scale with a sample of mothers with children under the age of 5 (Tsurikova, 2003, personal communication) received feedback that some participants did not feel the scale was applicable to their situation. One of the participants said that she could imagine men answering those questions, but not women. Another participant said that she felt the questions were dated and did not capture the current situation for families with young children.

- The ethnic/racial make-up of the sample was not reported. It would be useful to assess the scale's validity and reliability for multiple ethnic/racial groups.

Bibliography (studies that have used the measure)

Beutell, N. J., & Witting-Berman, U. (1999). Predictors of work-family conflict and satisfaction with family, job career and life. *Psychological Reports, 85*(3), 893-904.

Gutek, B. A., & Searle, S. (1991). Rational versus gender role explanations for work-family conflict. *Journal of Applied Psychology, 76*(4), 560-568.

Work Family/Work-Life Measures

TITLE OF MEASURE	INTERROLE CONFLICT SCALE
	Thomas, L. T., & Ganster, D. C. (1995). Impact of family-supportive work variables on work-family conflict and strain: A control perspective. *Journal of Applied Psychology, 80* (1), 6-15.
	Tsurikova, L. (2003). *Professional knowledge and work-family balance for women psychotherapists.* Masters Thesis, University of Massachusetts Lowell.
Contact Information	Richard E. Kopelman Baruch College The City University of New York 17 Lexington Avenue New York, NY 10010, USA e-mail: Richard_Kopelman@baruch.edu

Work Family/Work-Life Measures

TITLE OF MEASURE	WORK-FAMILY CONFLICT AND FAMILY-WORK CONFLICT SCALES
Source/Primary reference	Netemeyer, R., Boles, J., & McMurrian, R. (1996). Development and validation of Work-Family Conflict and Family-Work Conflict Scales. *Journal of Applied Psychology, 81*(4), 400-410.
Construct measured	Conflict generated in family life because of work, and conflict generated at work because of family
Brief description	The instrument has 10 items with two subscales (consisting of 5 items each): 1. work-to-family conflict 2. family-to-work conflict. Items are rated on a 7-point scale, where 1 = strongly disagree to 7 = strongly agree. The items were designed to measure the conflict itself versus the outcomes of work-family or family-work conflict.
Sample items	Work-to-Family Conflict (WFC) - The demands of my work interfere with my home and family life. - The amount of time my job takes up makes it difficult to fulfill family responsibilities. Family-to-Work Conflict (FWC) - The demands of my family or spouse/partner interfere with work-related activities. - Family-related strain interferes with my ability to perform job-related duties.
Appropriate for whom (i.e. which population/s)	Employed adults
Translations & cultural adaptations available	None known
How developed	The conceptual approach for this instrument is based on the premises that WFC and FWC are distinct but related forms of interrole conflict. Based on previous work, 110 items were generated to reflect the WFC and FWC concepts. Items include general, strain-based, and time-based WFC and FWC. Four faculty members evaluated each item and a variation of

Work Family/Work-Life Measures

TITLE OF MEASURE — **WORK-FAMILY CONFLICT AND FAMILY-WORK CONFLICT SCALES**

Cohen's kappa formula was used to choose the 43 items that would be retained.

To create the final version of the instrument, researchers used an iterative confirmatory procedure with LISREL VII. Some of the items were deleted based on:

- Low loading on the intended factor
- Within-factor correlated measurement error, across-factor correlated error, or both
- Standardized factor loadings
- Redundant wording with other items

The general demand, time-based, and strain-based items were carried over because they met the criteria for retention. After three iterations, five items for each scale were chosen for the instrument.

Psychometric properties

<u>STUDY SAMPLES</u>

Participants		Sample 1	Sample 2	Sample 3
Sample Size		$n = 182$	$n = 162$	$n = 186$
Description		Teachers	Small business owners	Real estate salespersons
Age	*Median*	43	45	48
Gender	*Female*	$n = 128$	†	$n = 142$
	Male	†	$n = 96$	†
Race/Ethnicity		Not reported	Not reported	Not reported
Marital Status	*Married*	$n = 157$	$n = 130$	$n = 148$
Have children living at home		$n = 93$	$n = 65$	$n = 60$

†Not reported

<u>VALIDITY</u>

Construct Validity

Factor analyses confirmed that the subscales are empirically distinct.

Concurrent Validity

All three samples completed other surveys beside the WFC and FWC scales. The researchers predicted negative correlations between organizational commitment, job satisfaction, life satisfaction, relationship agreement, relationship satisfaction, and WFC and FWC. Positive correlations were predicted for Maslach Burnout Inventory (MBI; Maslach & Jackson, 1981), job tension, role conflict, role ambiguity, intention to leave the organization, and search for another job. The tables show the correlation coefficients.

	Sample 1		Sample 2		Sample 3	
Measure	WFC	FWC	WFC	FWC	WFC	FWC
Organizational commitment	-.20*	-.25**				
Job satisfaction	-.36**	-.30**	-.21*	-.16*	-.27**	-.22**
MBI	.56**	.38**	.47**	.19*		
Job tension	.58**	.32**	.43**	.23*	.55**	.38**
Role conflict	.40**	.33**				
Role ambiguity	.39**	.35**				
Intention to leave an organization	.25**	.23**	.14	.02	.28**	.17*
Search for another job	.12	.18*	.19*	.04	.17*	.19**
Life satisfaction	-.33**	-.44**	-.41**	-.32**	-.53**	-.35**
Relationship satisfaction	-.01	-.16*	-.30**	-.26**	-.27**	-.20**
Relationship agreement	-.14*	-.29**	-.24*	-.20*		

$*p < .05; ** p < .01$

RELIABILITY

Internal Consistency

Internal consistency is provided by construct reliability, coefficient alpha, and average variance extracted estimates.

	WFC			FWC		
Sample	Construct $\alpha =$	Coefficient $\alpha =$	Average	Construct $\alpha =$	Coefficient $\alpha =$	Average
1	.88	.88	.60	.87	.86	.58
2	.89	.89	.60	.82	.83	.48
3	.88	.88	.59	.90	.89	.64

Comments The ethnic/racial make-up of the sample was not reported. It would be useful to assess its validity and reliability for multiple ethnic/racial groups.

Work Family/Work-Life Measures

TITLE OF MEASURE	WORK-FAMILY CONFLICT AND FAMILY-WORK CONFLICT SCALES
Bibliography (studies that have used the measure)	Aryee, S., Luk, V., & Leung, A. (1999). Role stressors, interrole conflict, and well-being: The moderating influence of spousal support and coping behaviors among employed parents in Hong Kong. *Journal of Vocational Behavior, 54*(2), 259-278. Burke, R. J., & Greenglass, E. R. (2001). Hospital restructuring, work-family conflict and psychological burnout among nursing staff. *Psychology & Health, 16*(5), 583-865.
Contact Information	Richard Netemeyer McIntire School of Commerce University of Virginia P.O. Box 400173 Charlottesville, VA 22904-4173, USA Tel: 434-924-3388 e-mail: rgn3p@forbes2.comm.virginia.edu

Work Family/Work-Life Measures

TITLE OF MEASURE	WORKER PERCEPTION OF WORK SPILLOVER
Source/Primary reference	Small S. A., & Riley D. (1990). Toward a multidimensional assessment of work spillover into family life. *Journal of Marriage and the Family, 52*, 51-61.
Construct measured	Spillover of work into home/personal life.
Brief description	The scale is a 20-item measure. Following the authors' multidimensional model, the measure of worker perception of work spillover consists of four separate role context subscales: 1. spillover into the marital relationship 2. spillover into the parent-child relationship 3. spillover into leisure time 4. spillover into household tasks Items are presented as declarative statements and respondents are asked to indicate their degree of agreement with each item on a 5-point scale ranging from 1 = strongly disagree to 5 = strongly agree.
Sample items	Marital relationship scale - My job helps me have a better relationship with my spouse. - Worrying about my job is interfering with my relationship with my spouse. Parent-child relationship - My job makes it hard for me to have a good relationship with my child(ren). - My working hours interfere with the amount of time I spend with my child(ren). Leisure - My job makes it difficult for me to enjoy my free time outside of work. - The amount of time I spend working interferes with how much free time I have. Home management - My job makes it difficult for me to get household chores done. - I spend so much time working that I am unable to get done at home.

Work Family/Work-Life Measures

TITLE OF MEASURE	WORKER PERCEPTION OF WORK SPILLOVER
Appropriate for whom (i.e. which population/s)	Working adults
Translations & cultural adaptations available	None known
How developed	This 20-item measure of work spillover was developed by the authors specifically for this study based on their knowledge of the phenomena and relevant literature. Each item was designed to ask about a causal relationship between work and home life, i.e., with work spillover as the cause and consequences for home life as the effect.
Psychometric properties	*See below*

Psychometric properties

STUDY SAMPLE

Participants		Demographics
Sample Size		$n = 130$
Description		Married male executives with children
Age	Mean	43.8
Gender	Male	100%
Race/Ethnicity		Nearly all were white
Hours worked per week (M)		49.4

RELIABILITY

The Cronbach's α coefficient was .93 for the overall 20-item work spillover measure (Small & Riley, 1990).

Aryee (1993) found the following reliabilities.

Subscale	α =
Job-spouse	.70
Job-parent	.81

Comments: The scale was developed with a white male sample. More research would be needed to assess its usefulness with female samples. It would be useful to assess the scale's validity and reliability for multiple ethnic/racial groups.

Bibliography (studies that have used the measure): Ayree, S. (1993). Dual-earner couples in Singapore: An examination of work and non-work sources of their experienced burnout. *Human Relations, 46*(12), 1441-1469.

Work Family/Work-Life Measures

TITLE OF MEASURE: **WORKER PERCEPTION OF WORK SPILLOVER**

Contact Information

Stephen Small
Human Development & Family Studies
University of Wisconsin
1300 Linden Drive
Madison, WI 53706, USA

Tel: 608-263-5688

e-mail: sasmall@facstaff.wisc.edu

Work Family/Work-Life Measures

TITLE OF MEASURE	WORK-TO-FAMILY CONFLICT
Source/Primary reference	Stephens, G. K., & Sommer, S. M. (1996). The measurement of work-to-family conflict. *Educational and Psychological Measurement, 56*(3), 475-486.
Construct measured	Extent to which work demands affect family
Brief description	This scale has explicit directionality and consists of three subscales based on three conceptual dimensions of work-family conflict: time, strain, and behavior. It includes 14 items rated on a 7-point scale from 1 = strongly disagree to 7 = strongly agree. 1. Time-based conflict (four items) that are related to competition for the individual's time 2. Strain-based conflict (four items) where stress from the work domain produces strain and/or difficulty managing both roles 3. Behavior-based conflict (six items) when patterns of behavior appropriate to each role are incompatible
Sample items	▪ My work keeps me from my family more than I would like. (time-based) ▪ I often feel the strain of attempting to balance my responsibilities at work and home. (strain-based) ▪ I am not able to act the same way at home as at work. (behavior-based)
Appropriate for whom (i.e. which population/s)	Working adults
Translations & cultural adaptations available	None
How developed	Twenty-eight items were developed from a review of the literature that addressed work-family conflict. They were classified according to 47 subject matter experts. With only two exceptions, only items that achieved 80% agreement among the experts were retained, leaving 16 items. The next step was an exploratory factor analysis to explore the factor structure of the measurement items. Fourteen items were retained based on the fact that they loaded on only one factor.

Work Family/Work-Life Measures

TITLE OF MEASURE: WORK-TO-FAMILY CONFLICT

Psychometric properties

STUDY SAMPLES

	Phase I	Phase II	
Participants	Sample 1	Sample 2	Sample 3
Sample Size	$n = 300$	$n = 145$	$n = 128$
Description	Employees of a large rehabilitation hospital	Employees of a large state service and regulatory agency	Employees of a contract diagnostic testing firm
Response rate	88-100%	91%	71%
Age (mean)	37	40	33.5
Gender — Female	87%	47%	66%
Gender — Male	13%	53%	34%
Race/Ethnicity	Not reported	Not reported	Not reported
Married	61%	81%	64%
Have Children	75%	78%	37%

VALIDITY

Construct Validity

Phase I: Traditional factor analysis with orthogonal rotation (varimax) in the first phase of the study resulted in three factors. The first one included 8 items that were originally constructed to measure time and strain dimensions. The other two dimensions included items designed to measure the behavioral domain of work-to-family conflict.

Phase II: Confirmatory factor analyses were performed and the three-factor solution was concluded to provide the best fit. The factors were similar to the domains of the original theoretical model.

Comments

- The measure is unique in the way it is grounded in an explicit directionality of conflict between work and family roles.

- Additional research is needed to establish the reliability and validity of the measure in general.

- The ethnic/racial make-up of the sample was not reported. It would be useful to assess the scale's validity and reliability for multiple ethnic/racial groups.

Bibliography (studies that have used the measure)

Stephens, G. K., & Sommer, S. M. (1995). Linking work-family conflict, work-based social support, and work group climate with job involvement and organizational citizenship behavior: Testing a path analytic model. *Journal of Health and Human Services Administration, 18*(1), 44-67.

Work Family/Work-Life Measures

TITLE OF MEASURE	WORK-TO-FAMILY CONFLICT
	Dawn Carlson, among others, has also used this instrument in some of her work.
Contact Information	Except for possible copyright costs from the journal, there are no costs that the author knows about for fair use of this instrument.
	Gregory K. Stephens, Ph.D. Associate Professor and Chair Department of Management M.J. Neeley School of Business Texas Christian University TCU Box 298530 Fort Worth, Texas 76129, USA
	Tel: 817-257-7548
	Fax: 817-257-7227
	e-mail: g.stephens@tcu.edu

References

References

Alexanderson K, Leijon M, Akerlind I, Rydh H, Bjurulk P [1994]. Epidemiology of sickness and absence in a Swedish county in 1985, 1986, and 1987. Scandinavian Journal of Social Medicine, *22*(1):27–34.

Allen TD, Herst DE, Bruck CS, Sutton M [2000]. Consequences associated with work-to-family conflict: A review and agenda for future research. Journal of Occupational Health Psychology, *5*(2):278–308.

Altemeyer B [1981]. Right-wing authoritarianism. Winnipeg, Canada: University of Manitoba Press.

Ayree S [1993]. Dual-earner couples in Singapore. An examination of work and non-work sources of their experienced burnout. Human Relations, *46*(12), pp. 1441–1469.

Bacharach SB, Bamberger P, Conley S [1991]. Work-home conflict among nurses and engineers. Mediating the impact of role stress on burnout and satisfaction at work. Journal of Organizational Behavior *12*(1):39–53.

Barbarin OA, Gilbert R [1981]. Institutional racism scale: Assessing self and organizational attributes. In: Barbarin OA, Good PR, Pharr OM, Sisking J, eds. Institutional Racism and Community Competence Rockville Maryland: U.S. Department of Health and Human Services. pp. 147–171.

Barnett RC [2003]. Community: The missing link in work-family research. Unpublished paper presented at the Workforce/Workplace Mismatch: Work, Family, Health, and Well-being conference, Washington, DC, June, 2003.

Barnett RC, Brennan RT. [1995]. The relationship between job experience and psychological distress. A structural equation approach. Journal of Organizational Behavior *16*:1–18.

Barnett RC, Gareis KC. [forthcoming]. Parental after-school stress and psychological well-being. Manuscript submitted for publication in Journal of Marriage and Family.

Barnett RC, Gareis KC, Brennan RT [1999]. Fit as a mediator of the relationship between work hours and burnout. Journal of Occupational Health Psychology *4*:307–317.

Barnett RC, Marshall NL, Raudenbush SW, Brennan RT [1993]. Gender and the relationship between job experiences and psychological distress. A study of dual-earner couples. Journal of Personality and Social Psychology, *64*:794–806.

Barling J, Dekker I, Laughlin C, Kelloway E, Fullagar C, Johnson D [1996]. Prediction and replication of the organizational and personal consequences of workplace sexual harassment. Journal of Managerial Psychology *11*(5):4–26.

Behson SJ [2002]. Coping with family-to-work conflict: The role of informal work accommodations to family. Journal of Occupational Health Psychology *7*(4):324–341.

References

Belkic K, Landsbergis P, Schnall P, Baker D [2004]. Is job strain a major source of cardiovascular disease risk? Scandinavian Journal of Work Environment & Health *30*:85–128.

Benson P, Vincent S [1980]. Development and validation of the Sexist Attitudes Toward Women Scale (SATWS). Psychology of Women Quarterly *5*:276–291.

Bergers GPA, Marcelissen FHG, Wolff Ch. J, de [1986]. Vos-D, vragenlijst organisatiestress-D, Handleiding (Vos-D, Questionnaire of Organizational Stress-D, Handbook). Nijmegen: Katholieke Universiteit Nijmegen.

Bianchi S, Milkie MA, Sayer LC, Robinson JP [2000]. Is anyone doing the housework? Trends in the gender division of household labor. Social Forces *79*(1):1–39.

Bingham SG, Scherer LL [1993]. Factors associated with responses to sexual harassment and satisfaction with outcome. Sex Roles *29*(3–4):239–270.

Blau FD [1998]. Trends in the well-being of American women. 1970–1995. Journal of Economic Literature *36*(1):112 165. pp. 152–153

Blau FD, Ferber M, Winkler A [2002]. The Economics of Women, Men, and Work. 4th Ed. Upper Saddle River, NJ: Prentice Hall.

Bohen H, Viveros-Long A [1981]. Balancing job and family life: Do flexible work schedules help? Philadelphia: Temple University Press.

Bond MA [1995]. Prevention and the ecology of sexual harassment: Creating empowering climates. Prevention in Human Services *12*(2):147–173.

Bond MA [2003] Prevention of sexual harassment. Gullotta InT Bloom M eds. Encyclopedia of primary prevention and health promotion, pp. 969–975. New York: Kluwer Publishing.

Borg V, Kristensen TS [2000]. Social class and self-rated health: Can the gradient be explained by differences in life style or work environment? Social Science and Medicine *51*(7):1019–1030.

Brigham JC [1993]. College students' racial attitudes. Journal of Applied and Social Psychology *23*:1933–1967.

Browne B [1997]. Gender and beliefs about work force discrimination in the United States and Australia. Journal of Social Psychology *137*(1):107–116.

Burden DS, Googins B [1987]. Boston University balancing job and homelife study: Managing work and family stress in corporations. Boston, MA: Boston University School of Social Work, Working Paper.

Burley K [1989]. Work-family conflict and marital adjustment in dual career couples: A comparison of three time models [Dissertation-doctoral]. Claremont, CA: Claremont Graduate School Unpublished.

References

Burt MR. [1980]. Cultural myths and support for rape. Journal of Personality and Social Psychology. *38*:217–230.

Büssing A [1996]. Social tolerance of working time scheduling in nursing. Work and Stress *10*(3):238–250.

Campbell A [1971]. White attitudes toward black people. Ann Arbor: Institute for Social Research.

Campbell A, Converse PE, Miller WE, Stokes DE [1960]. The American voter. New York: Wiley.

Cedillo-Becerril L [1999]. Psychosocial risk factors among women workers in the maquiladora industry in Mexico. [Dissertation Doctoral]. Lowell, MA: University of Massachusetts Lowell, Dept. of Work Environment.

Cervantes RC, Padilla AM, Salgado de Snyder VN [1987]. Development of the Latin American Stress Inventory (LAS-I). Unpublished.

Cervantes RC, Padilla AM, Salgado de Snyder VN. [1991]. The Hispanic Stress Inventory: A culturally relevant approach to psychosocial assessment. Journal of Consulting and Clinical Psychology *3*(3):438–447.

Clark R, Anderson NB, Clark V, Williams DR [1999]. Racism as a stressor for African Americans. American Psychologist *54*(10):805–816.

Cohen PN, Bianchi Suzanne M [1999]. Marriage, children, and women's employment: What do we know? Monthly Labor Review *122*(12):22–31.

Cohen S, Karmarck T, Mermelstein R [1983]. A global measure of perceived stress. Journal of Health and Social Behavior *24*:385–396.

Cortina LM [2001]. Assessing sexual harassment among Latinas: Development of an instrument. Cultural Diversity and Ethnic Minority Psychology *7*(2):164–181.

Crowne DP, Marlowe D [1960]. A new scale of social desirability independent of psychopathology. Journal of Consulting Psychology *24*:349–354.

Crowne D, Marlowe D [1964]. The approval motive. New York: Wiley.

Dansky B, Kilpatrick D [1997]. Effects of sexual harassment. In W. O'Donohue (Ed.), Sexual harassment: Theory, research, and treatment pp. 152–174. Boston: Allyn & Bacon.

Darity WA [2003]. Employment discrimination, segregation, and health. American Journal of Public Health *93*(2):226–231.

Dekker S, Schaufeli W [1995]. The effects of job insecurity on psychological health and withdrawal: A longitudinal study. Australian Psychologist *30*:57–63.

References

Derogatis LR [1977]. SCL-90 (Revised) version manual-I. Baltimore, MD: Johns Hopkins University School of Medicine.

Derogatis LR, Lipman RS, Rickles K, Uhlenhuth E, Covi L [1974]. The Hopkins Symptom Checklist: A self-report symptom inventory. Behavioral Science *19*:1 15.

Deutsch CJ [1985]. A survey of therapists' personal problems and treatment. Professional Psychology Research and Practice *16*(2):305 315.

Diaz RM, Ayala G, Bein E, Henne J, Marin B [2001]. The impact of homophobia, poverty, and racism on the mental health of gay and bisexual Latino men: Findings from 3 US cities. American Journal of Public Health *91*(6):927 932.

Dohrenwend BS, Krasnoff L, Askenasy AR, Dohrenwend BP. [1978]. Exemplification of a method for scaling life events: The PERI Life Events Scale. Journal of Health and Social Behavior *19*:205–229.

Dovidio JF Gaertner SL [1996]. Affirmative action, unintentional racial biases, and intergroup relations. Journal of Social Issues *52*(4):51–75.

Duleep HO [1986]. Measuring the effect of income on adult mortality. Journal of Human Resources *21*:238–251.

Dunton BC, Fazio RH [1997]. An individual difference measure of motivation to control prejudiced reactions. Personality and Social Psychology Bulletin *23*:316–326.

Eagly AH [1967]. Involvement as a determinant of response to favorable and unfavorable information. Journal of Personality and Social Psychology *7*(3):1–15.

Eagly AH, Mladinic A, Otto S [1991]. Are women evaluated more favorably than men? Psychology of Women Quarterly *15*:203–216.

Eaton SC [1999]. Gender and the structure of work in biotechnology. The Annals of the New York Academy of Science 865:175–188.

Eaton SC, Bailyn L [1999]. Emergent career paths in changing organizations: Work and life strategies of professionals in biotechnology firms. Annals of the American Academy of Political and Social Science *562*:159–173.

Essed P [1991]. Understanding everyday racism. Newbury Park, CA: Sage.

Evans KM, Herr EL [1991]. The influence of racism and sexism in the career development of African American women. Journal of Multicultural Counseling & Development *19*(30):130–136.

References

Feldman JJ, Makve DM, Kleinman JC, et al. [1989]. National trends in educational differences in mortality. American Journal of Epidemiology *129*:919–933.

Fitzgerald LF, Drasgow F, Hulin CL, Gelfand MJ, Magley VJ [1997]. Antecedents and consequences of sexual harassment in organizations: A test of an integrated model. Journal of Applied Psychology *82*(4):578 589.

Fitzgerald LF, Gelfand MJ, Drasgow F [1995]. Measuring sexual harassment: Theoretical and psychometric advances. Basic and Applied Social Psychology *17*(4): 425 445.

Fitzgerald LF, Ormerod AJ [1991]. Perceptions of sexual harassment: The influence of gender and context. Psychology of Women Quarterly *15*:281 294.

Friedman S Greenhaus J [2000]. Work and family: Allies or enemies. New York, NY. Oxford University Press.

Frone MR Russell M [1995]. Job stressors, job involvement and employee health: A test of identity theory. Journal of Occupational and Organizational Psychology *68*(1):1–12.

Frone MR Russell M [1997]. Relation of work-family conflict to health outcomes: A four-year longitudinal study of employed women. Journal of Occupational and Organizational Psychology *70*(4):325–336.

Frone MR, Russell M, Barnes GM [1996]. Work-family conflict, gender and health-related outcomes: A study of employed parents in two community samples. Journal of Occupational Health Psychology *1*:57–60.

Frone MR, Russell M, Cooper ML [1992a]. Antecedents and outcomes of work-family conflict: Testing a model of the work-family interface. Journal of Applied Psychology *77*:65–78.

Frone MR, Russell M, Cooper ML [1992b]. Prevalence of work-family conflict: Are work and family boundaries asymmetrically permeable? Journal of Organizational Behavior *13*:723–729.

Frone M, Yardley JK [1996]. Workplace family-supportive programs: Predictors of employed parent's importance ratings. Journal of Occupational and Organizational Psychology *69*(4):351–367.

Gaertner SL, Dovidio JF [1986]. The aversive form of racism. In JF Dovidio, SL Gaertner eds. Prejudice, discrimination and racism. Orlando, FL: Academic Press. pp. 61–89

Galinsky E, Bond J, Friedman D [1993]. The changing workforce: Highlights of the national study. New York: Families and Work Institute.

Gareis KC, Barnett RC [2001] Schedule fit and stress-related outcomes among women doctors with families. Unpublished paper presented at the annual meeting of the American Psychological Association, San Francisco, CA, August, 2001.

References

Geurts S, Taris T, Kompier et al. (forthcoming). SWING: Development and validation of the Survey Work-home Interaction-Nijmegen' in five different occupational groups. Available from Sabine Geurts at S.Geurts@psych.kun.nl.

Glick P, Fiske ST [1996]. The ambivalent sexism inventory: Differentiating hostile and benevolent sexism. Journal of Personality and Social Psychology 70(3):491–512.

Glick P, Fiske ST [1999]. The ambivalence toward men inventory. Psychology of Women Quarterly 23(3):519–536.

Glick P et al. [2000]. Beyond prejudice as simple antipathy: Hostile and benevolent sexism across cultures. Journal of Personality and Social Psychology 79:763–775.

Glick P et al. [2003]. Hostile as well as Benevolent Attitudes Toward Men Predict Gender Hierarchy: A 16-Nation Study. Lawrence University, Appleton, WI. Unpublished.

Glomb TM, Richman WL, Hulin CL, Drasgow F, Schneider KT, & Fitzgerald LF [1997]. Ambient sexual harassment: An integrated model of antecedents and consequences. Organizational Behavior and Human Decision Processes 71(3):309–328.

Godfrey S, Richman C, Withers T [2000]. Reliability and validity of a new scale to measure prejudice: The GRISMS. Current Psychology 19(1):8–13.

Goldenhar L, Swanson N, Hurrell J, Ruder A, Deddens J [1998]. Stressors and adverse outcomes for female construction workers. Journal of Occupational Health Psychology 3(1):19–32.

Goldsmith E [1989]. Work involvement, family involvement, role overload, and fatigue of employed men and women. Florida State University, College of Home Economics ,Tallahassee, FL: Working Paper.

Green NL [1995]. Development of the Perceptions of Racism Scale Image: Journal of Nursing Scholarship 27(2):141–146.

Gutek BA [2001]. Women and paid work. Psychology of Women Quarterly 25(4):379–393.

Gutek BA, Done R [2001]. Sexual harassment. In: R Unger ed., Handbook of the psychology of women and gender. New York: Wiley pp. 367–387.

Gutek BA, Koss MP [1993]. Changed women and changed organizations: Consequences of and coping with sexual harassment. Journal of Vocational Behavior 42:28–48.

Gutek BA, Searle S, Klepa L [1991]. Rational versus gender role explanations for work-family conflict. Journal of Applied Psychology 76:560–568.

References

Hall EM [1992]. Double exposure: The combined impact of the home and work environments on psychosomatic strain in Swedish women and men. International Journal of Health Services *22*:239–260.

Hanisch KA [1990]. A causal model of general attitudes, work withdrawal, and job withdrawal, including retirement. [Unpublished doctoral dissertation]. Champaign, IL: University of Illinois at Urbana.

Hanisch KA [1996]. An integrated framework for studying the outcomes of sexual harassment: Consequences for individuals and organizations. In M. S. Stockdale ed., Sexual Harassment in the workplace: Perspectives, frontiers, and response strategies: Vol. 5. Thousand Oaks, CA: Sage. pp. 174 198.

Hanisch KA, Hulin CL [1990]. Job attitudes and organizational withdrawal: An examination of retirement and other voluntary withdrawal behaviors. Journal of Vocational Behavior *37*:60 78.

Hanisch KA, Hulin CL [1991]. General attitudes and organizational withdrawal: An evaluation of a causal model. Journal of Vocational Behavior *39*:110–128.

Harrell SP [1994]. The Racism and Life Experience Scale—Revised. Unpublished manuscript.

Harrell SP [1997]. Researching racism and psychological well-being: Conceptualization and change. Paper presented at the 105th Convention of the American Psychological Association, Chicago, IL 1997.

Harrell SP [2000]. A multidimensional conceptualization of racism-related stress: Implications for the well-being of people of color. American Journal of Orthopsychiatry *70*:42–57.

Harrell SP, Merchant MA, Young SA [1997]. Psychometric properties of the Racism and Life Experiences Scales (RaLES). Presented at the annual meeting of the American Psychological Association. Chicago, IL, August, 1997.

Hayghe HV, Bianchi SM [1994]. Married mothers' work patterns: The job-family compromise. Monthly Labor Review *117*(6):24–30.

Hesson-McInnis MS, Fitzgerald LF [1997]. Sexual harassment: A preliminary test of an integrative model. Journal of Applied Social Psychology *27*(10):877–901.

Hudson WW, Ricketts WA [1980]. A strategy for the measurement of homophobia. Journal of Homosexuality *5*:357–372.

Huebner D [2002]. Mental and physical health consequences of perceived discrimination. [Unpublished doctoral dissertation]. Arizona State University.

References

Hughes D, Dodge MA [1997]. African American women in the workplace: Relationships between job conditions, racial bias at work, and perceived job quality. American Journal of Community Psychology 25(5):581–599.

Hughes D, Galinski E [1994]. Gender, job and family conditions, and psychological symptoms. Psychology of Women Quarterly 18(2):251–271.

Hughes D, Johnson D [2001]. Correlates in children's experiences of parents' racial socialization behaviors. Journal of Marriage and Family 63(4):981–996.

Hulin C, Fitzgerald L, Drasgow F [1996]. Organizational Influences on sexual harassment, In M. Stockdale, ed., Sexual Harassment in the Workplace: Perspectives, frontiers and response strategies. Thousand Oaks, CA: Sage.

Ironson G [1992]. Work, job stress and health. In: S. Zedeck ed. Work, Families and Organizations. San Francisco, CA: Jossey-Bass.

Iverson RD, Olekalns M, Erwin PJ [1998]. Affectivity, organizational stressors, and absenteeism: A causal model of burnout and its consequences. Journal of Vocational Behavior 52:1–23.

Jackson DNA [1970]. A sequential system for personality scale development. Current Topics in Clinical and Community Psychology 2:61–96.

Jackson JS, Brown TN, Williams DR, Torres M, Sellers SL, Brown K [1996]. Racism and the physical and mental health status of African Americans: A thirteen-year national panel study. Health, Ethnicity, & Disease 6(1–2):132–147.

Jacobson CR [1985]. Resistance to affirmative action: Self-interest or racism. Journal of Conflict Resolution 29:306–329.

Jeanquart-Barone S, Sekaran U [1996] Institutional racism: An empirical study. Journal of Social Psychology 136(4):477–482.

Joosten J, Drop MJ [1987]. De betrouwbaarheid en vergelijkbaarheid van drie versies van de VOEG. Gezondheid En Samenleving.

Josephson M, Pernold G, Ahlberg-Hultén G, Härenstam A, Theorell T, Vingård E et al. [1999]. Differences in the association between psychosocial work conditions and physical workload in female- and male-dominated occupations. American Industrial Hygiene Association Journal 60:673–678.

Kanner AD, Coyne JC, Schaeffer C, Lazarus RS [1981]. Comparison of two modes of stress measurement: Daily hassles and uplifts vs. major life events. Journal of Behavioral Medicine 4:1–39.

Karasek RA [1985]. Job Content Instrument: Questionnaire and user's guide. Los Angeles, CA: Department of Industrial and Systems Engineering.

References

Karasek RA, Theorell T [1990]. Healthy work: Stress, productivity and the reconstruction of working life. New York: Basic Books.

Katz I, Hass RG [1988]. Racial ambivalence and American value conflict: Correlational and priming studies of dual cognitive structures. Journal of Personality and Social Psychology *55*:893–905.

Kessler RC, Mickelson KD, Williams DR [1999]. The prevalence, distribution, and mental health correlates of perceived discrimination in the United States. Journal of Health and Social Behavior *40*:208–230.

Klitzman S, House JS, Israel BA, Mero RP [1990]. Work stress, nonwork stress, and health. Journal of Behavioral Medicine *13*(3):221–243.

Klonoff EA, Landrine H [1995]. The schedule of sexist events. Psychology of Women Quarterly *19*(4):430–472.

Klonoff EA, Landrine H, Campbell R [2000]. Sexist discrimination may account for well-known gender differences in psychiatric symptoms. Psychology of Women Quarterly *24*(1):93–99.

Klonoff EA, Landrine H, Ullman JB [1999]. Racial discrimination and psychiatric symptoms among blacks. Cultural Diversity and Ethnic Minority Psychology *5*(4):329–339.

Komarovsky M [1977]. Dilemmas of masculinity. New York: Norton.

Kopelman RE, Greenhaus JH, Connoly TF [1983]. A model of work, family, and interrole conflict: A construct validation study. Organizational Behavior and Human Performance *32*(2):198–215.

Korabik K, McDonald LM, Rosin HM [1993]. Stress, coping, and social support among women managers. In: Long BC, Kahn SE eds. Women, work, and coping. A multidisciplinary approach to workplace stress Montreal & Kingston, Canada: McGill-Queen's University Press. pp. 133–153.

Kossek E, Ozeki C [1998]. Work-family conflict, policies, and the job-life satisfaction relationship: A review and directions for future organizational behavior-human resources research. Journal of Applied Psychology *83*:139–149.

Kossek EE, Zonia SC [1993]. Assessing diversity climate: A field study of reactions to employer efforts to promote diversity. Journal of Organizational Behavior *14*:61–81.

Krieger N [1990]. Racial and gender discrimination: Risk factors for high blood pressure? Social Science Medicine *30*(12):1273–1281.

References

Krieger N [1995] What Explains the Public's Health? - A call for epidemiologic theory. Epidemiology 7:107–109.

Krieger N [1999] Embodying Inequality: A review of concepts, measures, and methods for studying health consequences of discrimination. International Journal of Health Services 29(2):295–352.

Krieger N [2003]. Does racism harm health? Did child abuse exist before 1962? On explicit questions, critical science and current controversies: An ecosocial perspective. American Journal of Public Health 93(2):194–199.

Krieger N, Rowley DL, Herman AA, Avery B, Phillips MT [1993]. Racism, sexism, and social class: Implications for studies of health, disease, and well-being. American Journal of Preventive Medicine 9:82–122.

Krieger N, Sidney S [1996]. Racial discrimination and blood pressure: The cardia study of young black and white adults. American Journal of Public Health 10:1370–1378.

Krieger N, Sidney S [1997]. Prevalence and health implications of anti-gay discrimination: A study of black and white women. International Journal of Health Services 27(1):157–176.

Kwate NO, Valdimarsdottir HB, Guevarra JS, Bovbjerg DH [2003]. Experiences of racist events are associated with negative health consequences for African American women. Journal of the National Medical Association 95(6):450–460.

Landrine H, Klonoff EA [1994]. The African American Acculturation Scale: Development, reliability, validity. Journal of Black Psychology 20:104–127.

Landrine H, Klonoff EA [1996]. The Schedule of Racist Events: A measure of racial discrimination and a study of its negative physical and mental health consequences. Journal of Black Psychology 22(2):144–169.

Landrine H, Klonoff EA, Gibbs J, Manning V, Lund M [1995]. Physical and psychiatric correlates of gender discrimination. An application of the Schedule of Sexist Events. Psychology of Women Quarterly 19:473–492.

Lane R. (1955). Four-item F scale in "political personality and electoral choice." American Political Science Review 49:173–190.

Larkey LK [1996]. The development and validation of the Workforce Diversity Questionnaire: An instrument to assess interactions in diverse workgroups. Management Communication Quarterly 9(3):296–338.

Larsen KS, Reed M, Hoffman S [1980]. Attitudes of heterosexuals toward homosexuality: A Likert-type scale and construct validity. Journal of Sex Research 16:245–257.

References

Lazarus RS, Folkman S [1984]. Stress, appraisal, and coping. New York: Springer.

Leary MR [1983a]. A brief version of the Fear of Negative Evaluation Scale. Personality and Social Psychology Bulletin 9:371–375.

Leary MR [1983b]. Social anxiousness: The construct and its measurement. Journal of Personality Assessment 47:66–75.

Lenhart S [1996]. Physical and mental health aspects of sexual harassment. In: Shrier DK ed., Sexual harassment in the workplace and the academia. Psychiatric issues. Vol. 38.Washington DC: American Psychiatric Press, Inc., pp. 21–38.

Levins R, Lopez C [1999]. Toward an ecosocial view of health. International Journal of Health Services 29(2):261–293.

Loscocco KA [1997]. Work-family linkages among self-employed women and men. Journal of Vocational Behavior 50:204–226.

Lott B [1995]. Distancing from women: Interpersonal sexist discrimination. In: Lott B, Maluso D eds. The social psychology of interpersonal discrimination. New York: The Guilford Press, pp. 13 49.

Marmot M, Smith GD, Stansfeld S et al. [1989]. Health inequalities among British civil servants: the Whitehall II study. Lancet 337:1387–1393.

Marmot M [1999]. Importance of the psychosocial environment in epidemiologic studies. Scandinavian Journal of Work Environmental Health 25:49–53.

Maslach C, Jackson SE [1981]. The measurement of experienced burnout. Journal of Occupational Behavior 2:99–113.

Matthews S, Hertzman C, Ostry A, Power C [1998]. Gender, work roles and psychosocial work characteristics as determinants of health. Social Science and Medicine 46:1417–1424.

Mays VM, Coleman LM, Jackson JS [1996]. Perceived race-based discrimination, employment status, and job stress in a national sample of black women: Implications for health outcomes. Journal of Occupational Health Psychology 1(3):319–329.

McConahay JB [1986]. Modern racism, ambivalence, and the modern racism scale. In: Dovidio J, Gaertner S eds. Prejudice, discrimination and racism. San Diego: Academic Press, pp. 91–125

McConahay JB, Hardee BB, Batts V [1981]. Has racism declined? It depends upon who's asking and what is asked. Journal of Conflict Resolution 25:563–579.

McConahay JB, Hough JC [1976]. Symbolic racism. Journal of Social Issues 32, pp. 23–45.

References

McNeilly MD, Anderson NB, Armstead CA, Clark R, Corbett M, Robinson EL, Pieper CF, Lepisto EM [1996]. The Perceived Racism Scale: A multidimensional assessment of the experience of white racism among African Americans. Health, Ethnicity and Disease 6:154–166.

Meijman TF, Mulder G [1998]. Psychological aspects of workload. In: Drenth PJ, Thierry H, De Wolff CJ eds. Handbook of work and organizational psychology, 2nd ed. Hove, England UK: Psychology Press/Erlbaum (Uk) Taylor & Francis, pp. 5–33

Messick D, Mackie DM [1989]. Intergroup relations. Annual Review of Psychology 40:45–81.

Messing K [1995]. Chicken or egg: Biological differences and the sexual division of labor. In: Messing K, Neis B, Dumais L eds. Invisible: Issues in women's occupational health/La santé des travailleuses. Charlottetown, P.E.I., Canada, Gynergy Books. pp. 177–201.

Messing K [1997]. Women's occupational health: A critical review and discussion of current issues. Women and Health 25:39–68

Meyer IH [2003]. Prejudice as stress: Conceptual and measurement problems. American Journal of Public Health 93(2):262–265.

Mirels H, Garrett J [1971]. The protestant ethic as a personality variable. Journal of Consulting and Clinical Psychology 36:40–44.

Murdoch M, McGovern PG [1998]. Measuring sexual harassment: Development and validation of the Sexual Harassment Inventory. Violence and Victims 13(3):203–216.

Murrell AJ, Olson JE, Hanson-Frieze I [1995]. Sexual harassment and gender discrimination: A longitudinal study of women managers. Journal of Social Issues 51(1):139–149.

Nazroo JY [2003]. The structuring of ethnic inequalities in health: Economic position, racial discrimination, and racism. American Journal of Public Health 93(2):277–84.

Neighbors HW, Jackson JS, Broman C, Thompson E [1996]. Racism and the mental health of African Americans: The role of self and system blame. Health, Ethnicity, & Disease, 6(1–2) 167–175.

Nelson DL, Quick JC, Hitt MA [1989]. Men and women of the personnel profession: Some differences and similarities in their stress. Stress Medicine 5:145–152.

Netemeyer R, Boles J, McMurrian R [1996]. Development and validation of Work-Family Conflict and Family-Work Conflict Scales. Journal of Applied Psychology 81(4):400–410.

References

Nordander C, Ohlsson K, Balogh I, Rylander L, Pålsson B, Skerfving S [1999]. Fish processing work: The impact of two sex dependent exposure profiles on musculoskeletal health. Occupational and Environmental Medicine 56:256–264.

Parker SK, Griffin MA [2002]. What is so bad about a little name-calling? Negative consequences of gender harassment for overperformance demands and distress. Journal of Occupational Health Psychology 7(3):195–210.

Paulhus DL [1988]. Assessing self-deception and impression management in self-reports: The Balanced Inventory of Desirable Responding. University of British Columbia, Vancouver, British Columbia, Canada.

Pavalko E, Mossakowski K, Hamilton V [2003]. Does perceived discrimination affect health? Longitudinal relationships between work discrimination and women's physical and emotional health. Journal of Health and Social Behavior 44(1):18–33.

Peeters MCW [1994]. Supportive interactions and stressful events at work: An event-recording approach. Proefschrift, Katholieke Universiteit Nijmegen.

Pinel EC [1999] Stigma consciousness: The psychological legacy of social stereotypes. Journal of Personality and Social Psychology 76(1):114–128.

Piotrkowski CS [1998]. Gender harassment, job satisfaction, and distress among employed white and minority women. Journal of Occupational Health Psychology 3:33–43.

Piotrkowski CS [1987]. Work and the Family System: A Naturalistic Study of Working-Class and Lower Middle-Class Families. New York: Free Press.

Plant EA, Devine PG [1998]. Internal and external motivation to respond without prejudice. Journal of Personality and Social Psychology 75(3):811–832.

Pogrebin MR, Poole ED [1997]. The sexualized work environment: A look at women jail officers. Prison Journal 77(1):41–57.

Ponterotto JG, Burkard A, Rieger BP [1995]. Development and initial validation of the Quick Discrimination Index (QDI). Educational and Psychological Measurement 55:1016–1031.

Ponterotto JG, Reiger BP, Barrett A, Harris G, Sparks R, Sanchez CM, Magids D [1993]. Development and initial validation of the Multicultural Counseling Awareness Scale (MCAS). Paper presented at the Ninth Buros-Nebraska Symposium on Measurement and Testing: Multicultural Assessment, Lincoln, NE, 1993.

Punnett L, Herbert R [2000]. Work-related musculoskeletal disorders: is there a gender differential, and if so, what does it mean? In: Goldman MB, Hatch M eds. Women and Health. San Diego, CA: Academic Press, pp. 474–492

References

Putney S, Middleton R [1961]. Dimensions and correlates of religious ideologies. Social Forces *39*:285–290.

Quinn MM, Woskie SR, Rosenberg BJ [2000]. Women and work. In: Levy BS, Wegman DH eds. Occupational health: Recognizing and preventing work-related disease and injury. 4th ed. Philadelphia, PA.: Lippincott, Williams & Wilkins, pp. 655–678

Radloff LS [1977]. The CES-D scale: A self report depression scale for research in the general population. Applied Psychological Measurement *1*:385–401.

Repetti RL, Matthews KA, Waldron I [1989]. Employment and women's health: Effects of paid employment on women's mental and physical health. The American Psychologist *44*:1394–1401.

Richman JA, Rospenda KM, Nawyn SJ, Flaherty JA, Fendrich M, Drum ML et al. [1999]. Sexual harassment and generalized workplace abuse among university employees: Prevalence and mental health correlates. American Journal of Public Health *89*(3):358–363.

Riger S, Stokes J, Raja S, Sullivan M [1997]. Measuring perceptions of the work environment for female faculty. The Review of Higher Education *21*(1):63–78.

Rosenberg M [1965]. Society and the adolescent self-image. Princeton, NJ: Princeton University Press.

Rothausen TJ [1999]. "Family" in organizational research: A review and comparison of definitions and measures. Journal of Organizational Behavior *20*:817–836.

Rowley DL [1994]. Research issues in the study of very low birth weight and preterm delivery among African Americans. Journal of National Medical Association *86*(10):761–764.

Salgado de Snyder VN [1987]. Factors associated with acculturative stress and depressive symptomatology among married Mexican immigrant women. Psychology of Women Quarterly *11*:475–488.

Schnall PL, Landsbergis PA, Baker D [1994]. Job strain and cardiovascular disease. Annual Review of Public Health *15*:381–411.

Schneider KT, Swan S, Fitzgerald LF [1997]. Job-related and psychological effects of sexual harassment in the workplace: Empirical evidence from two organizations. Journal of Applied Psychology *8*(3):223–233.

Schulz A, Israel B, Williams D, Parker E, Becker E, Becker A, James S [2000]. Social inequalities, stressors and self reported health status among African American and white women in the Detroit metropolitan area. Social Science and Medicine *51*:1639–1653.

References

Schuman H, Harding J [1963]. Sympathetic identification with the underdog. Public Opinion Quarterly 27:230–241.

Sears DO [1988]. Symbolic Racism. In: Katz PA, Taylor DA eds. Eliminating racism: Profiles in controversy. New York: Plenum Press, pp. 53–84

Shrier DK ed. [1996]. Sexual harassment in the workplace and the academia: Psychiatric issues Vol. 38, Washington, DC: American Psychiatric Press, Inc., pp. 1–20

Siegrist J [1996]. Adverse health effects of high-effort/low-reward conditions. Journal of Occupational Health Psychology 1:27–41.

Small SA, Riley D [1990]. Toward a multidimensional assessment of work spillover into family life. Journal of Marriage and the Family 52:51–61.

Smith PC, Kendall L, Hulin CL [1969]. The measurement of satisfaction in work and retirement. Chicago: Rand McNally.

Snyder M, Gangestad S [1986]. On the nature of self-monitoring: Matters of assessment, matters of validity. Journal of Personality and Social Psychology 51:125–139.

Spence JT, Helmreich R [1972]. The Attitudes Toward Women Scale: An objective instrument to measure attitudes toward the rights and roles of women in contemporary society. JSAS Catalog of Selected Documents in Psychology 2:66–67 (Ms. No. 153).

Spence JT, Helmreich R, Holahan CK [1979]. Negative and positive components of psychological masculinity and femininity and their relationships to self-reports of neurotic and acting out behaviors. Journal of Personality and Social Psychology 37:1673–1682.

Spence JT, Helmreich R, Stapp J [1973]. A short version of the attitudes toward women scale (AWS). Bulletin of the Psychonomic Society 2:219–220.

Spence JT, Helmreich R, Stapp J [1974]. The Personal Attributes Questionnaire: A measure of sex-role stereotypes and masculinity-femininity. JSAS Catalog of Selected Documents in Psychology 4:43(Ms. No. 617).

Stephens GK, Sommer SM [1996]. The measurement of work to family conflict. Educational and Psychological Measurement 56(3):475–486.

Stokes J, Riger S, Sullivan M [1995]. Measuring perceptions of the working environment for women in corporate settings. Psychology of Women Quarterly 19(4):533–549.

Swanson N [2000]. Working women and stress. Journal of the American Medical Women's Association 55(2):76–79.

Swim JK, Aikin KJ, Hall WS, Hunter BA [1995]. Sexism and racism: Old-fashioned and modern prejudices. Journal of Personality and Social Psychology 68:199–214.

References

Swim JK, Hyers LL, Cohen LL, Ferguson MJ [2001]. Everyday sexism: Evidence for incidence, nature, and psychological impact from three daily diary studies. Journal of Social Issues 57(1):31–53.

Tavris C [1973]. Who likes women's liberation—and why: The case of the unliberated liberals. Journal of Social Issues 29:175–198.

Terrell F, Miller FS [1980]. The development of an inventory to measure experience with racialistic incidents among blacks. Unpublished manuscript.

Terrell F, Terrell S [1981]. An inventory to measure cultural mistrust among blacks. The Western Journal of Black Studies 5(3):180–185.

Thomas LT, Ganster DC [1995]. Impact of family supportive work variables on work-family conflict and strain: A control perspective. Journal of Applied Psychology 80:6–15.

Thompson CE, Neville H, Weathers PL, Poston WC, Atkinson DR [1990]. Cultural mistrust and racism reaction among African American students. Journal of College Student Development 31:162–168.

Tougas F, Brown R, Beaton AM, Joly S [1995]. Neosexism: Plus ca change, plus c'est pareil. Personality and Social Psychology Bulletin 21(8):842–891.

Utsey SO, Chae M, Brown C, Kelly D [2002]. Effect of ethnic group membership on ethnic identity, race-related stress, and quality of life. Cultural Diversity and Ethnic Minority Psychology 8(4):366–377.

Utsey SO, Ponterotto JG [1996]. Development and validation of the Index of Race-Related Stress (IRRS). Journal of Counseling Psychology 43(4):490–501.

Van Veldhoven M, Meijman TF, Broersen JP, Fortuin RJ [1997]. Handleiding VBBA: Onderzoek naar de beleving van psychosociale arbeidsbelasting en werkstress met behulp van de vragenlijst beleving en beoordeling van de arbeid [Guideline VBBA: Research on the experience of psychosocial work load and job stress with the Questionnaire on the Experience and Evaluation of Work]. Amsterdam: SKB (Stichting Kwaliteitsbevordering BGZ Nederland).

Vercoulen JHM, Alberts M, Bleijenberg G [1999]. De Checklist Individuele Spankracht (CIS) The Checklist Individual Strength (CIS). Gedragstherapie 32:131–136.

Vermeulen M, Mustard C [2000]. Gender differences in job strain, social support at work, and psychological distress. Journal of Occupational Health Psychology 5:428–440.

Wagena E, Geurts SAE [2000]. SWING: ontwikkeling en validering van de 'Survey Werk-thuis Interferentie-Nijmegen. Gedrag en Gezondheid 28:138–158.

References

Waldo CR [1999]. Working in a majority context: A structural model of heterosexism as minority stress in the workplace. Journal of Counseling Psychology 46(2):218–232.

Walsh WB, Betz NE [1990]. Tests and assessment 2nd ed. Englewood Cliffs, NJ: Prentice Hall.

Warp P [1990]. The measurement of well-being and other aspects of mental health. Journal of Occupational Psychology 63:193–210.

Watson D, Friend R [1969]. Measurement of social-evaluative anxiety. Journal of Consulting and Clinical Psychology 33:448–457.

Watts RJ, Carter RT [1991]. Psychological aspects of racism in organizations. Group & Organization Management 16(3):328–345.

Weber L, Higginbotham E [1997]. Black and white professional-managerial women's perceptions of racism and sexism in the workplace. In: Higginbotham E., Romero M eds. Women and Work. Exploring race, ethnicity, and class. Thousand Oaks, CA: Sage.

Wethington E, Kessler RC [1989]. Employment, parental responsibility, and psychological distress: A longitudinal study of married women. Journal of Family Issues 10:527–546.

Williams DR, Neighbors HW, Jackson JS [2003]. Racial/ethnic discrimination and health: Findings from community studies. American Journal of Public Health 93(2):200–208.

Williams DR, Yu Y, Jackson JS, Anderson NB [1997]. Racial differences in physical and mental health: Socio-economic status, stress, and discrimination. Journal of Health Psychology 2(3):335–351.

Wooton BH [1997]. Gender differences in occupational employment. Monthly Labor Review 120(4):15–24.

Xu W, Leffler A [1996]. Gender and race effects on occupational prestige, segregation, and earnings. In: E. Ngan-Ling Chow, Wilkinson D, Baca Zinn M eds. Race, Class, & Gender: Common bonds, different voices. Thousand Oaks, CA: Sage.